THE PLASTIC ANISOTROPY IN SINGLE CRYSTALS AND POLYCRYSTALLINE METALS

The Plastic Anisotropy in Single Crystals and Polycrystalline Metals

by

WOJCIECH TRUSZKOWSKI

Institute of Metallurgy and Material Science,
Polish Academy of Sciences, Krakow, Poland

KLUWER ACADEMIC PUBLISHERS
DORDRECHT / BOSTON / LONDON

A C.I.P. Catalogue record for this book is available from the Library of Congress.

ISBN 978-90-481-5662-7

Published by Kluwer Academic Publishers,
P.O. Box 17, 3300 AA Dordrecht, The Netherlands.

Sold and distributed in North, Central and South America
by Kluwer Academic Publishers,
101 Philip Drive, Norwell, MA 02061, U.S.A.

In all other countries, sold and distributed
by Kluwer Academic Publishers,
P.O. Box 322, 3300 AH Dordrecht, The Netherlands.

Printed on acid-free paper

Published in cooperation with Polish Scientific
Publishers PWN

Contents

Acknowledgements

The author wishes to thank Dr. Jacenty Kloch from the Institute of Mathematics of the Polish Academy of Sciences for his keen interest and stimulating discussions during the preparation of this book.

This Monograph has been partially supported by the Committee for Scientific Research (KBN) under grant № 7 T08D 041 12.

Preface

The reader shall find in the offered monograph a systematic presentation of scientific effects in the field of anisotropy studies reached by the author and his collaborators in the period of recent four decades: published and discussed in a number of papers and conference contributions. The central construction line of discussion is to be sought in the full and comprehensive analysis of $r(\varepsilon)$ function defining the anisotropy coefficient varying during the tensile test. No doubt, this function can be considered as a nutshell carrier of comprehensive information about the essential features influencing the directionality of the studied material's plasticity. The function also provides the basis to elaborate methods used in the determination of such characteristics.

In the historical presentation of literature in the field of plastic anisotropy, the original input was offered by W.M. Baldwin Jr., already in 1946, who observed the differentiated strain rates in three mutually perpendicular directions of the sample subjected to static tensile test. In the following years, further and expanded analysis of the problem was undertaken by Lankford, Hill, Gensamer, Jackson, Low and Smith. However, the breakthrough in analytical approach to this problem was made by Krupkowski and Kawiński, as well as Lankford, Snyder and Bauscher, who proposed a quantitative method to define the strain ratio as the measure of plastic anisotropy. These contributions allowed a new interpretation of anisotropy relying on texture as the main reason and the drawability as one of its effects. At the same time, technological progress was made in the area of deep drawing of mild steel, which was an important contribution to the accelerated development of automobile industry.

In the fifties and sixties, research in this field was continued mainly in Europe: in Germany, Poland, U.K., and France, as well as a few American centres. The objective was defined mainly as the quantitative definition of crystallographic texture and its relation to plastic anisotropy. Yet, it was not until mid seventies that the rapid growth of car manufacturing in Japan inspired local research and development units to intensified studies over the relation between texture and the technology of plastic metal working, not only in connection to deep drawn steel. At the Sixth International Congress of Textures of Materials (ICOTOM-6) in Tokio, in 1981, a third of all papers were written and discussed by Japanese.

Today, we witness the publication of plastic anisotropy related studies in many countries of the world. The most important of them have been those which contributed to the definition of strict dependence between the developing directionality of physical properties and material characteristics. These findings enabled the subsequent development of research methods aiming at the comprehensive definition of the metal's susceptibility to drawing. Other publications have offered experimental data that could be used in further numerical analysis and technology development of plastic metal working, and are therefore practically useful, although they have not contributed much to the very concept of directionality of plastic properties in metals.

The monograph aims at a systematic presentation of the issue of emerging directionality, mainly in effect of the formation of crystallographic texture. This problem is related to the phenomenon of asymmetric scatter along the $r(\varepsilon)$ function both in polycrystalline metals and single crystals, which has been studied by the author and his collaborators in the recent decades. The obtained solution allowed to define a comprehensive characteristics of plastic anisotropy with determined physical sense, and the proposition of a fully quantitative description method.

For many years, the author has been also engaged in the study of many aspects of deformation of single crystals with defects of crystallographic orientation; the results of research justify the admission of the notion of "single crystal texture". One of the evident effects of crystallographic orientation defects lies in the instability of both crystallographic orientation and plastic anisotropy coefficient, in uniaxial tensile test. It can be recognized and defined quantitatively relying on the interpretation of $r(\varepsilon)$ function. The problem is of importance as the instability of texture and anisotropy parameters plays a significant role in the development of deformation texture. The proposed method provides also the possibility to survey the macroinhomogeneity of deformation of real crystals.

The majority of experimental results is given in the form of diagrams; however, some of them which, in the author's opinion could be used for founding another interpretation of the displayed relations are presented also in tables of numerical values.

Finally, it must be stressed that it was not the author's intention to offer an academic manual to lecturers or students of materials engineering or to engineers at metal works. The monograph is meant to be of help to those engaged in research and technology, suggesting other useful methods and effective procedures.

Chapter 1

Introduction

The following are the most important causes of the plastic anisotropy: anisotropy of individual crystals in polycrystalline metals, directionality of the distribution of particular phases and defects in multiphase alloys, leading to fibrous structure, and internal stresses developing in effect of oriented deformation. The latter factor is insignificant in recrystallized metals, while the texture, i.e. privileged crystallographic orientation, is the sole significant cause in the case of annealed single phase alloys (i.e. pure metals and solid solutions).

The paramount impact of plastic anisotropy in the technology of plastic working consists in the role played by texture and anisotropy in the process of metal sheet drawing.

The deep drawability of metals is most frequently defined on the basis of the strain hardening exponent n and strain ratio r, determined through interpretation of the effects of uniaxial tensile test. To this purpose, the stress – strain relationship $(\sigma - \varepsilon)$ is usually applied, defined by the mathematical formula showing: 1° best possible fit of experimental data, and 2° precise physical meaning of constant parameters. These two conditions constitute the criterion of the evaluation of $\sigma - \varepsilon$ relation from the viewpoint of its suitability to identify the strengthening mechanism of deformed metal.

It is worth mentioning that the determination of the strain hardening coefficient is sometimes difficult, because of inhomogeneous deformation at the tensile test.

This type of inhomogeneous deformation takes place when different mechanisms are launched in successive phases of the tensile test. It may happen that the deformation starts on one slip system, passes to another, and then to the region where twinning or shear banding predominate, different regions being described by different strain hardening exponents. The determination of the limit of the first zone is essential for correct description of the relations $\sigma(\varepsilon)$ and $r(\varepsilon)$, and the determination of the original value r_0, by means of the back extrapolation. This type of inhomogeneity is quite frequent. When it is neglected, the interpretation of experiments is erroneous, as will be shown in Chapter 2 [2.6 – 2.10, 2.12].

The difficulties that arise in the course of estimating the plastic anisotropy coefficient r with clearly defined physical meaning emerge from the variation of r in the course of straining. In effect, the calculation of the precise value of r_0, through the back extrapolation of $r(\varepsilon)$ function up to the initial state ($\varepsilon = 0$), becomes troublesome. The above function reflects the impact of straining texture upon the starting anisotropy being related to the type and and texture level of the studied material. In general, it has not been possible yet to determine it theoretically, relying on the parameters of a polycrystalline material.

The basic method used in the present study relies on the observation of crystallographic orientation changes in metals and alloys, through the analysis of the $r(\varepsilon)$ function. This is possible because the crystallographic orientation constitutes the only factor essentially influencing plastic anisotropy in annealed single phase alloys. Such procedure is fully justified in the evaluation of the instability of crystallographic orientation in nominally stable single crystals (with orientation symmetrical, in relation to the sample axis, in uniaxial tensile test). However, this method yields reliable effects only in the case when the coefficient of plastic anisotropy (strain ratio) clearly defines the material condition which is independent of the changes occurring in the test itself. Since r coefficient is usually measured after a significant deformation, the searched value of r_0 is defined through back extrapolation of the function $r = f(\varepsilon)$. If we do not know the form of function $r(\varepsilon)$, the found value of r_0 ($r_0 = r(0)$) must be indirectly reestablished from the viewpoint of physical interpretation. Moreover, the empirically defined function is characterized by significant experimental scatter. The scatter is asymmetrical and grows fast, parallel to the diminishing strain ε. All the above factors cause that the precise r_0 values found on the ground of experimental data can be determined only when proper mathematical method has been designed. It is helpful that much experimental material has been gathered, which allows to propose the general form of $r = f(\varepsilon)$ function.

The quantitative description of plastic anisotropy relying on the concept of strain ratio was developed independently in Poland, by Krupkowski and Kawiński [1.1], and in the United States, by Lankford, Snyder and Bauscher [1.2]. In both these essentially similar methods the measure of plastic anisotropy is K (Krupkowski) or r (Lankford). The value of both variables is defined in a tensile test usually at considerable strain. It means that the authors had assumed the invariability of anisotropy coefficients, or – alternatively – neglected as insignificant its variation with sample elongation, and justified such postulate by observations made on deep drawing mild steel. It was not until late fifties and the sixties that a number of papers were published where the authors attempted to demonstrate experimentally strain dependence of the strain ratio in *f.c.c.* and *b.c.c.* metals and alloys [1.3 – 1.39]. The objective of several papers was also the determination of the relationship between plastic anisotropy and crystallographic orientation, having in view the calculation of the *r* value in <uvw> single crystals as well as textured polycrystalline metals [1.40 – 1.45].

The problem of instability of crystallographic orientation in deformed single crystals, which can be scrutinized – according to the author's proposition – by means of $r(\varepsilon)$ function, has been related to the defects of crystal orientation: the deviation of sample axis from the assumed crystallographic axis, and the degree of "single crystal texture".

As shown above, the plastic anisotropy coefficient (strain ratio) is determined on the basis of correct interpretation of the tensile test. That is why Chapter Two contains the discussion of the stress – strain relationship both in the case of homogeneous and inhomogeneous deformations.

It should be mentioned here that Yoshida et al. [1.46] observed that *r* values did not always correlate with deep drawability in several metallic materials. They proposed the χ value as the stress ratio dependence in the Ludwik [2.15] equation with $\sigma_0 = 0$ (sometimes called Hollomon [2.18] equation). However, this proposition has not been widely recognized, and is not referred to in literature.

Chapters Three through Eight deal with the strain ratio as the measure of plastic anisotropy both in polycrystalline metals and in real single crystals.

In Chapter Nine, the author shows how the $r = f(\varepsilon)$ relationship of a nominally stable single crystal can be used in the formulation of the problem of instability of crystallographic orientation at uniaxial deformation.

The concluding Chapter Ten briefly discusses the role of plastic anisotropy described by the $r(\varepsilon)$ function in the research and the technology of metal working.

1.1   References

1.1. A.Krupkowski and S.Kawiński, *The Phenomenon of Anisotropy in Annealed Polycrystalline Metals*, J. Inst. Metals, **75**, 869 (1949).

1.2. W.T.Lankford, S.C.Snyder and J.A.Bauscher, *New Criteria for Predicting the Press Performance of Deep Drawing Sheets*, Trans. Am. Soc. for Metals, **42**, 1197 (1950).

1.3. W.Truszkowski, *Zagadnienie anizotropii zgniecionych metali polikrystalicznych*, Arch. Hutn., **1**, 171 (1956).

1.4. W.Truszkowski et Z.Bojarski, *Sur l'anisotropie de l'acier inoxydable 18/8 laminé à froid*, Mém. Sci. Rev. Métallurg., **59**, 112 (1962).

1.5. W.T.Roberts, *Crystallographic Aspects of Directionality in Sheet*, Sheet Met. Ind., **39**, 855 (1962).

1.6. S.Nagashima, H.Takechi and H.Kato, *The Plastic Strain Ratio of Polycrystalline Low Carbon Steel*, Trans. Japan Inst. of Metals, **5**, 274 (1964).

1.7. M.Atkinson and M.McLean, *The Measurement of Normal Plastic Anisotropy in Sheet Metals*, Sheet Met. Ind., **42**, 290 (1965).

1.8. W.T.Roberts, *Texture Control in Sheet Metal*, Sheet Met. Ind., **43**, 237 (1966).

1.9. M.Atkinson, *Assessing Normal Anisotropic Plasticity of Sheet Metals*, Sheet Met. Ind., **44**, 167 (1967).

1.10. R.L.Whiteley, *How Crystallographic Texture Controls Drawability*, Metal Progress, 81 (1968).

1.11. W.Truszkowski and J.Jarominek, *Plastic Anisotropy of Cold Rolled Copper*, Arch. Hutn., **14**, 309 (1969).

1.12. W.Truszkowski et J.Król, *Sur l'évaluation quantitative de l'anisotropie des métaux*, C.R. Acad. Sci. Paris, **269**, 807 (1969).

1.13. W.Truszkowski and J.Król, *A Quantitative Analysis of Plastic Anisotropy in Polycrystalline Copper*, Bull. Acad. Pol. Sci., sér. techn., **17**, 981 (1969).

1.14. S.Nagashima, H.Takechi and H.Kato, *Relation between Texture and r–Value in Steel Sheets*, Textures in Research and Practice, Proc. Intern. Symposium, Clausthal–Zellerfeld, 444 (1969).

1.15. W.Truszkowski et J.Król, *Tendance à l'anisotropie des métaux cubiques à faces centrées laminés à froid*, Mém. Sci. Rev. Métallurg., **67**, 201 (1970).

1.16. S.R.Goodman and Hsun Hu, *Effect of Hot Rolling Texture on the Plastic Strain Ratio of Low Carbon Steels*, Met. Trans., **1**, 1629 (1970).

1.17. J.S.Kallend and G.J.Davies, *Prediction of Plastic Anisotropy in Annealed Sheets of Copper and α – Brass*, Journ. Inst. Metals, **98**, 242 (1970).

1.18. G.J.Davies, D.J.Goodwill and J.S.Kallend, *Elastic and Plastic Anisotropy in Sheets of Cubic Metals*, Met. Trans., **3**, 1627 (1972).

1.19. W.Truszkowski et J.Jarominek, *Sur le sens physique de la méthode d'appréciation quantitative de l'anisotropie plastique*, C.R.Acad. Sci. Paris, **274**, 2053 (1972).

1.20. W.Truszkowski et J.Jarominek, *Essai de synthèse des recherches sur l'anisotropie plastique*, Mém. Sci. Rev. Métallurg., **70**, 433 (1973).

1.21. P.Parnière et G.Pomey, *Relations entre l'anisotropie cristallographique et l'anisotropie mécanique. Cas des tôles minces d'acier extra doux pour emboutissage*, Mécanique, Matériaux, Electricité, **XI**, 1 (1974).

1.22. W.Truszkowski, *On the Quantitative Evaluation of Plastic Anisotropy in Sheet Metals*, Proc. 8 th Biennial Congress of IDDRG., Gothenburg, 48 (1974).

1.23. Hsun Hu, *The Strain–Dependence of Plastic Strain Ratio (r_m value) of Deep Drawing Sheet Steels Determined by Simple Tension Test*, Met. Trans. A, **6A**, 945 (1975).

1.24. W.Truszkowski, *Influence of Strain on the Plastic Strain Ratio in Cubic Metals*, Met. Trans. A, **7A**, 327 (1976).

1.25. J.Kuśnierz et Z.Jasieński, *Courbe de traction et valeur propre du coefficient d'anisotropie des tôles d'aluminium, de cuivre et de laiton*, Mém. Sci. Rev. Mét., **73**, 485 (1976).

1.26. H.J.Bunge, D.Grzesik, G.Ahrndt and M.Schulze, *The relation between preferred orientation and the Lankford parameter r of plastic anisotropy*, Arch. Eisenhüttenwes., **52**, 407 (1981).

1.27. W.Truszkowski, S.Wierzbiński and A.Modrzejewski, *Influence of Mosaic Structure on Instability of the Strain Ratio in Deformed Copper Single Crystals*, Bull. Acad. Pol. Sci., sér. techn., **30**, 367 (1982).

1.28. W.Truszkowski, S.Wierzbiński, *Izmenenie koefficienta plastičeskoj anizotropii monokristallov medi s orientacjej blizkoj* [001] *pri rastjazenii*, Fizika Metallov i Metallovedenie, **56**, 1195 (1983).

1.29. W.Truszkowski, *Quantitative Aspects of the Relation between Texture and Plastic Anisotropy*, Proc. Intern. Conf. ICOTOM 7, Noordwijkerhout, 723 (1984).

1.30. W.Truszkowski et J.Kloch, *Modifications du coefficient d'anisotropie plastique en cours de déformation hétérogène*, Matériaux et Techniques, **E** 17 (1985).

1.31. W.Truszkowski, *On Proper Criteria of Correlating Texture and Plastic Anisotropy*, Bull. Pol. Ac.: Techn., **33**, 43 (1985).

1.32. W.Truszkowski, S.Wierzbiński and A.Modrzejewski, *Effet des défauts d'orientation cristallographique sur l'instabilité plastique des monocristaux de cuivre*, Arch. Hutn., **31**, 129 (1986).

1.33. J.Kloch, W.Truszkowski, *The Method of Determination of the Fitting Function Based on Maximal Errors*, Bull. Pol. Ac.: Techn., **34**, 683 (1986).

1.34. W.Truszkowski, J.Kloch, *Application of the Maximal Error Method for the Calculation of the $r(\varepsilon)$ Function*, Bull. Pol. Ac.: Techn., **34**, 691 (1986).

1.35. W.Truszkowski, S.Wierzbiński, A.Modrzejewski, J.Baczyński, G.S.Burchanov, I.V.Burov, and O.D.Čistjakov, *Influence of Deviation from* <001>, <011> *and* <111> *Orientations on the Variation of Strain Ratio in Deformed Nickel Single Crystals*, Arch. of Metallurgy, **32**, 165 (1987).

1.36. W.Truszkowski, *The Impact of Texture in Single Crystals of FCC Metals on Mechanical Behaviour and Instability of Orientation*, Proc. 8 th Intern. Conf. ICOTOM-8, Santa Fe, 537 (1988).

1.37. W.Truszkowski, S.Wierzbiński, J.Kloch, *Količestvennyj analiz plastičeskoj anizotropii polikristalličeskoj medi*, Fiz. Met. Metalloved., **66**, 178 (1988).

1.38. W.Truszkowski, A.Modrzejewski and J.Baczyński, *Variation of the Strain Ratio in Tensile Tested* <011> *Brass Single Crystals*, Bull. Pol. Ac.: Techn., **37**, 471 (1989).

1.39. W.Truszkowski and A.Modrzejewski, *Influence of Stacking Fault Energy on Instability of Crystallographic Orientation in Tensile Tested Brass Single Crystals*, Arch. of Metallurgy, **35**, 219 (1990).

1.40. A.Krupkowski, *Anizotropia mono- i polikrystalicznego metalu o strukturze A1*, Arch. Hutn., **2**, 9 (1957).

1.41. J.A.Elias, R.H.Heyer and J.A.Smith, *Plastic Anisotropy of Cold Rolled – Annealed Low Carbon Steel Related to Crystallographic Orientation*, Trans Met. Soc. AIME, **224**, 678 (1962).

1.42. W.F.Hosford, *Discussion of the paper of Elias, Heyer and Smith*, Trans. Met. Soc. AIME, **227**, 272 (1963).

1.43. W.Truszkowski, J.Gryziecki and J.Jarominek, *Assessment of the Strain Ratio in the Cube Plane of f.c.c. Metals*, Bull. Acad. Pol. Sci., sér. techn., **24**, 209 (1976).

1.44. W.Truszkowski, J.Gryziecki and J.Jarominek, *Variation of Strain Ratio in Cube Plane of Copper*, Metals Technology, **6**, 439 (1979).

1.45. W.Truszkowski, J.Gryziecki and J.Jarominek, *Variation of Plastic Strain Ratio in the* {001} *Crystallographic Plane of Silver*, Bull. Pol. Ac.: Tech., **31**, 31 (1983).

1.46. Kiyota Yoshida, Koichi Yoshii, Hiroshi Komorida, Matsuo Usuda and Hajime Watanabe, *Significance of the X value and the work-hardening exponent* [n] *under equibiaxial tension in the assessment of sheet metal formability*, Sheet. Met. Ind., **48**, 772 (1971).

Chapter 2

Work hardening of the material at the uniaxial tensile test

2.1 Introduction

The description of work hardening relationship (σ stress $-\varepsilon$ strain) through mathematical function (e.g. formulae 2.1. or 2.7.) allows to define the course of homogeneous deformation in the range where constant and unchanging deformation mechanism is active. The function parameters, which adopt constant values in the entire range of homogeneous deformation, have a precise physical meaning. For example, the coefficient m defines the material's strain – hardening ability, while the change of mechanism corresponds to the shift to another range, having different m value. It is in this sense that the term of macroinhomogeneous deformation is used: a clear delimitation of zones occurs here in relation to stress – strain dependence, caused by the operation of different deformation mechanisms.

Zankl [2.1], and later a number of authors: Schwink and Knoppik [2.2], Krause and Götler [2.3], Essman et al. [2.4], relying on experiments carried out on high stacking fault energy polycrystalline *f.c.c.* metals (mainly copper and nickel), have identified four intervals at the stress – strain curve. In the zone of very small strains (below $\varepsilon = 0.001$) accommodation processes are observed. Afterwards, Zone I begins (up to $\varepsilon = 0.01$), where multiple slip occurs in the largest grains. This zone is essentially different from the area of easy glide in single crystals. In contrast, polycrystalline zones II and III are governed by a mechanism basically similar to zone II and III of single crystals. In Zone II, easy glide predominates in one system, while multiple slip is active only in areas close to grain surface. The appearance of zone III results from the development of the cross slip mechanism.

Already in the fifties, Crussard and Jaoul [2.5 and 2.6] found that the relations $\sigma(\varepsilon)$, in the samples of aluminium, copper, and aluminium alloys, cannot be defined by Ludwik's formula ($\sigma = \sigma_0 + K\varepsilon^n$), with constant parameters σ_0, K and n, within the entire straining range. The change of parameters corresponds to the boundaries between zones of linear course of the function $\log \sigma = f(\log \varepsilon)$.

In the case of *f.c.c.* metals with low stacking fault energy, as well as in hexagonal *h.c.p.* metals, in a situation of a limited number of slip systems, one can expect also the activation of the twinning process and the emergence of a new range on the $(\sigma - \varepsilon)$ curve.

The development of new zones at the stress – strain curve in low stacking fault energy copper alloys with tin and zinc, as well as in cobalt and nickel, was observed by Krishnamurthy et al. [2.7]. In their study on the effect of temperature and strain rate upon the mechanism of the deformation of poly-crystalline technical purity titanium, Truszkowski, Łatkowski and Dziadoń [2.8] identified in coarse-grain material (with grain size of 170 μm) ranges, where the slip mechanism predominated in prismatic and pyramidal systems, as well as the zones where the reduced *m* coefficient value had been caused by twinning. The increased temperature and stress resulted in the development of a new deformation mechanism, and a new interval at the $\sigma - \varepsilon$ curve, where the higher strain – hardening coefficient was caused by dynamic ageing.

As the stages of a polycrystalline specimen are much less evident than those of a single crystal, Reed–Hill et al. [2.9] recommended a special form of analysis to reveal them. The material macroinhomogeneity may also be a good cause of troubles in distinguishing the boundaries of zones.

The methods of the determination of stages in the stress – strain relationship have been proposed by Styczyński and Jasieński [2.10], and Grabianowski et al. [2.11].

2.2   The stress – strain relationship for homogeneous deformation, defined by power function

The issue of adopting appropriate mathematical formula to define stress – strain relationship in the case of uniaxial straining has been the subject of much discussion for many years. A number of critical synthetic studies [2.12 – 2.14] clearly indicate the preponderance of the power-type formulae.

The following are most frequently used: Ludwik's formula (1909) [2.15] and Krupkowski's equation (1946) [2.16], later formulated also by Swift (1952) in a slightly modified version [2.17]. Hollomon's formula (1945) [2.18] is but a simplified version of Ludwik's equation. Its applicability is limited [2.19].

Relying on the criterion of stability loss in a tensile test, and experimental results obtained in research over homogeneously deformed metals, such as copper, mild steel and 18/8 type austenitic steel, Truszkowski [2.12, 2.14] has shown that Krupkowski–Swift equation coefficients have clear physical interpretation.

If σ is the true stress, and ε is the natural strain, Ludwik's formula can be written as follows:

$$\sigma = \sigma_0 + K\varepsilon^n \tag{2.1}$$

In order to define the instantaneous value of the n coefficient, with no need to determine σ_0, Crussard and Jaoul [2.5, 2.6] proposed the differential form of equation (2.1):

$$\frac{d\sigma}{d\varepsilon} = Kn\varepsilon^{n-1} \tag{2.2}$$

Having introduced the following symbol:

$$\xi = \frac{d\sigma}{d\varepsilon} \tag{2.3}$$

the straight linear relationship between $\log \xi$ and $\log \varepsilon$ is obtained:

$$\log\xi = (n-1)\log\varepsilon + \log(Kn) \tag{2.4}$$

In 1946, Krupkowski proposed the following formula:

$$\sigma = k z_i^m \tag{2.5}$$

where:

$$z_i = z_1 + (1 - z_1) z \tag{2.6}$$

and z (cold work) is the relative reduction in the sample cross section.

When natural strain (rather than cold work) is applied as the deformation measure, Krupkowski's equation adopts the form proposed by Swift (1952), i.e.:

$$\sigma = k(\varepsilon_0 + \varepsilon)^m \tag{2.7}$$

Thus, in the double logarithmic system, the following relationships are obtained:

$$\log\xi = \frac{m-1}{m}\log\sigma + \log(km) \qquad\qquad (2.8)$$

$$\log\xi = (m-1)\log(\varepsilon_0 + \varepsilon) + \log(km) \qquad\qquad (2.9)$$

The formulae (2.4) as well as (2.8) and (2.9) allow to define the tensile curve parameters relying on experimental data interpreted according to Ludwik $\xi = f(\varepsilon)$, or Swift $\xi = f(\sigma)$ or $\xi = f(\varepsilon_0 + \varepsilon)$. The coefficient ξ_i is the slope of the curve $\sigma = f(\varepsilon)$ at the point $(\varepsilon_i, \sigma_i)$.

As pointed out by Kocks et al. [2.20], Reed–Hill, Cribb and Monteiro [2.9] claim that the values of ξ and σ are parameters defined by the material condition (density and distribution of dislocations), rather than its past history, as it is the case in relation to ε. Thus the parameters m, k, and ε_0 have more essential physical interpretation, compared to n, K and σ_0. All arguments point to the statement that the most comprehensive information about the studied material and its plastic deformability is obtained from the presentation of the tensile test results in the form of the relationship $\xi = f(\sigma)$, rather than $\xi = f(\varepsilon)$. A similar opinion has been voiced also by other authors [2.3, 2.9, 2.20]. The finding of the relative advantage of stress – strain presentation relying on $\xi(\sigma)$ over the relationship $\xi(\varepsilon)$ is tantamount to the assertion of Krupkowski–Swift formula superiority over Ludwik's equation. This conclusion fully corresponds to the author's thesis, proved in earlier publications [2.12, 2.14, 2.19].

2.3   The characteristics of macroinhomogeneous deformation

In consequence of the above findings, the macroinhomogeneous deformation will be defined below with the aid of Krupkowski–Swift formula with constant parameters in each zone (though different in particular zones). The borders between strain zones are the less evident, the more macroinhomogeneous is the studied material.

Krishnamurthy, Qian and Reed–Hill [2.7] studied the impact of mechanical twinning upon the development of stages in the uniaxial stress – strain curve at room temperature. Low stacking fault energy alloys were studied: Cu 3.1% at Sn,   Cu 4.9% at Sn,   Cu 30% at Zn,   and   Co 40% at Ni.   The authors found that twinning was largely responsible for the formation of four ranges at the strain hardening curve. In the first range easy glide predominates; in the second one – twins develop in particular crystals, at the same crystallographic plane; the third range is characterized by the twinning at intersecting planes. In the fourth range, the dynamic recovery becomes the dominating process. In the second range, the twinning leads to low value of the *m* strain hardening coefficient, while the intersecting twins in the third range result in the growth of the mentioned coefficient, along with the clear reduction of grain size. These results are reflected in the diagram for Cu 4.9% at Sn (Fig. 2.1).

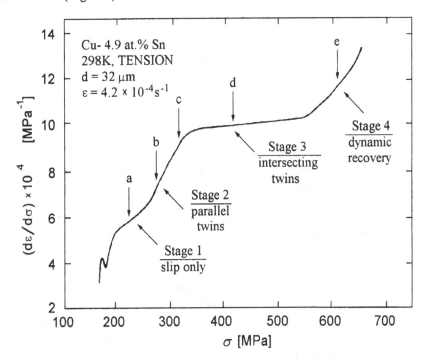

Figure 2.1. The development of inverse *dε/dσ* hardening coefficient in relation to *σ* in Cu-4.9 Sn alloy at the temperature 298 K [2.7]

In order to study the development of the function $\xi(\sigma)$ and $\xi(\varepsilon_0 + \varepsilon)$ in conditions of the macroinhomogeneous deformation of strained coarse-grained titanium, the experimental results obtained by Truszkowski, Łatkowski and Dziadoń [2.8] were used. Figure 2.2 shows the three-stage diagrams presented according to the formula (2.9) for technical purity tita-nium strained at the rate $\dot{\varepsilon} = 1.5 \, 10^{-2} \, \text{sec}^{-1}$ at room temperature, and at temperatures of 100 °C and 200 °C.

At room temperature, the diagram $\xi(\varepsilon_0 + \varepsilon)$, similarly as the diagram $\xi(\sigma)$, indicate three distinct ranges. In the first one, easy glide is the predominant deformation method. However, in the case of certain grains unfavourably oriented for slip, deformation twinning is observed. The high value of the *m*

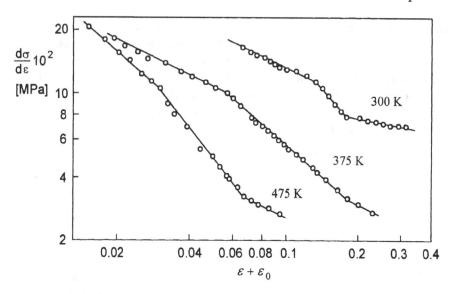

Figure 2.2. The relationship $d\sigma/d\varepsilon = f(\varepsilon_0 + \varepsilon)$ for coarse-grained titanium (170 μm) deformed by straining at different temperatures, at the rate $\dot{\varepsilon} = 1.5 \ 10^{-2} \ sec^{-1}$, [2.8]

coefficient in this range can be explained as a result of the slip–slip, or slip–twinning relationship. The second range, accompanied by rapid decline of the strain hardening coefficient, is caused by intensive twinning. It can be assumed that the third range, showing the *m* value significantly higher compared to range one, relies basically on the slip in newly oriented grains, in result of twinning, as well as the twinning in cross-secting planes.

In conclusion, the stress – strain relationship in macrohomogeneous deformation is better defined by Krupkowki–Swift formula, assuming initial deformation ε_0, than by Ludwik equation, relying on the initial stress σ_0. In the case of macroinhomogeneous deformation, which exhibits distinct ranges at the strain hardening curve dominated by different physical mechanisms, the definition of $\sigma(\varepsilon)$ function must be limited to the studied range. Thus, the $\sigma - \varepsilon$ relationship in the entire deformation area shall be defined by Krupkowski – Swift formula, where the constant parameters adopt different values in particular ranges.

Several papers on strain hardening regimes in tension and compression tests have been published in recent years [2.21, 2.22]. In their paper on low stacking fault energy *f.c.c.* alloys Asgari , El-Danaf, Kalindindi and Doherty [2.23] present a sort of synthesis of conclusions reported in the literature.

2.4 References

2.1. G.Zankl, Z. Naturforschung, **18a**, 795 (1963).

2.2. Ch.Schwink und D.Knoppik, *Experimentelle Untersuchungen zum plastischen Verhalten von vielkristallinen Nickel*, Phys. Stat. Sol., **8**, 729 (1965).

2.3. D.Krause und E.Götler, *Untersuchungen zur Verfestigungskurve vielkristallinen Kupfers*, Phys. Stat. Sol., **9**, 485 (1966).

2.4. U.Essmann, M.Rapp und M.Wilkens, *Die Versetsungsanordnung in plastisch verformten Kupfer-Vielkristallen*, Acta Met., **16**, 1275 (1968).

2.5. Ch.Crussard et B.Jaoul, *Contribution à l'étude de la forme des courbes de traction des métaux et à son interprétation physique*, Rev. de Métallurgie, **47**, 589 (1950).

2.6. Ch.Crussard, *Rapport entre la forme exacte des courbes de traction des métaux et les modifications concomitantes de leur structure*, Rev. de Métallurgie, **50**, 697 (1953).

2.7. S.Krishnamurthy, K.W.Qian and R.E.Reed-Hill, *Effects of Deformation Twinning on the Stress – Strain of Low Stacking Fault Energy Face-Centered Cubic Alloys*, Practical Applications of Quantitative Metallography, ASTM STP 839, Philadelphia, 41 (1984).

2.8. W.Truszkowski, A.Łatkowski and A.Dziadoń, *Stress – Strain Behaviour and Microstucture of Polycrystalline Alpha-Titanium*, Proc. 2nd Risø Intern. Symp. on Metallurgy and Mat. Science, 383 (1981).

2.9. R.E.Reed-Hill, W.R.Cribb and S.Monteiro, *Concerning the Analysis of Tensile Stress – Strain Data Using log dσ/dε$_p$ Versus log σ Diagrams*, Met. Trans., **4**, 2665 (1973).

2.10. A.Styczyński, Z.Jasieński, *Zakresy odkształcenia w próbie rozciągania polikrystalicznej miedzi*, Arch. Hutn., **15**, 309 (1970).

2.11. A.Grabianowski, J.Kloch and W.Kuczek, *Determination of the Strain Hardening Relationship of Metals Based on Error Rectangles*, Bull. Ac. Pol. Sci., sér. techn., **24**, 831 (1976).

2.12. W.Truszkowski, *Analiza procesu deformacji w próbie rozciągania przy uwzględnieniu niejednorodności metali*, Arch. Hutn., **4**, 283 (1959).

2.13. H.J.Kleemola and M.A.Nieminen, *On the Strain Hardening Parameters of Metals*, Met. Trans., **5**, 1863 (1974).

2.14. W.Truszkowski, *On Physical Meaning of the Stress – Strain Relationship Parameters in High Strength Polycrystalline Metals*, Mém. Sci. Rev. Métallurg., **77**, 193 (1980).

2.15. P.Ludwik, *Elemente der technologischen Mechanik*, J.Springer, 32, (1909).

2.16. A.Krupkowski, *The Deformation of Plastic Metals by Strain*, Annales de l'Acad. Pol. des Sc. Techn., **7**, 113 (1946).

2.17. H.W.Swift, *Plastic Instability Under Plane Stress*, J. Mech. Phys. Solids, **1**, 1 (1952).

2.18. J.H.Hollomon, *Tensile Deformation*, Trans. AIME, **162**, 268 (1945).

2.19. W.Truszkowski, *On the Usefulness of Hollomon Equation for the Interpretation of the Stress – Strain Relationship of Titanium*, Arch. Hutn., **26**, 395 (1981).

2.20. U.F.Kocks, H.S.Chen, D.A.Rigney and R.J.Shaefer, *Work Hardening*, AIME Met. Conf. Series, **46**, 151, J.P.Hirth and Weertman, eds., Gordon and Breach Sc. Publ., New York, 1968 (cit. in [2.9]).

2.21. Kiyota Yoshida, Koichi Yoshii, Hiroshi Komorida, Matsuo Usuda and Hajime Watanabe, *Significance of the X value and the work-hardening exponent [n] under equibiaxial tension in the assessment of sheet metal formability*, Sheet. Met. Ind., **48**, 772 (1971).

2.22. W.Truszkowski, *Divers aspects de l'hétérogénéité induite par déformation dans les alliages à faible énergie des fautes d'empilement*, Archives of Metallurgy, **38**, 139 (1993).

2.23. Sirous Asgari, Ehab El-Danaf, Surya R. Kalindindi and Roger D. Doherty, *Strain Hardening Regimes and Microstructural Evolution during Large Strain Compression of Low Stacking Fault Energy Fcc Alloys That Form Deformation Twins*, Met. and Mat. Trans., **28A**, 1781 (1997).

Chapter 3

Strain ratio as the measure of plastic anisotropy

3.1   Introduction

Each stage in the production of metal sheet, i.e. casting, plastic working and heat treatment leads to the formation of preferred crystallographic orientation. The texture is an important (though not the exclusive) cause of anisotropy of physical properties, including plastic properties of metals and alloys. Anisotropy can be observed in metals submitted to some procedure. The anisotropy effects – permanent or transient – are easily found in every crystal. It may also be the consequence of the uniform distribution of the second phase, or impurities and segregation products following the plastic working. When the function defining the impact of internal stresses on a certain physical variable is not linear, it is possible that internal stresses cause anisotropy. Finally, the neighbouring crystals in a polycrystalline sample interact in the tension test; the effect of such interaction depends on their mutual crystallographic orientation.

Plastic anisotropy is an important factor influencing the drawability of metal sheets. Two types of plastic anisotropy can be identified. The differentiated elongation propensity in the sheet plane and in the normal direction (n o r m a l   a n i s o t r o p y) determines the drawability of sheets, while the p l a n a r   a n i s o t r o p y – reflected by different strain ratio values in various directions of sheet plane – results in the formation of ears in the drawn cups, as well as puckering and uneven thinning.

The advantageous effect of plastic anisotropy on formability of sheets has been generally accepted in science and technology only recently. Although, in 1950, Lankford and collaborators showed the influence of normal anisotropy upon the drawability of steel sheets, even in the sixties it was generally believed that anisotropy was related to unfavourable impact upon the plastic working of metals. It was deemed that only in special cases it could be expected that planar anisotropy could make the technological process more efficient. Such approach can be found even in the third edition (1966) of the crucial textbook [Barrett and Massalski – 3.1] where it is claimed that:

"An example in which anisotropy is often undesirable is the steel sheet used for deep drawing. Unequal mechanical properties caused by a specific texture in a sheet may cause difficulties or waste in certain deep-drawn operations. Thus, a cup drawn from such a sheet may have an uneven rim or «ears». On the other hand, anisotropy may be put to advantage if the drawing operation itself is unsymmetrical and if the sheet can be turned to a position that makes favourable use of the directionality."

Today, it is widely accepted that the correctly used directionality of mechanical properties favours the technological process of plastic working, while some unfavourable aspects of directionality can be prevented through the application of various known methods.

By 1949, the measurement methods of plastic anisotropy were very rudimentary, and no clear interpretation of results could be offered. Usually, the tensile test was effected upon the samples cut from the sheet at various angles to the rolling direction, and the following mechanical properties were measured: yield strength, tensile strength, uniform elongation, total elongation, reduction in area. The image of plastic anisotropy obtained in this way had certain applicability in the assessment of non-symmetrical elements pressing. Some other technological methods were also suggested, such as the tear test proposed by Brownsdon [3.2] and Jevons [3.3], though the results obtained provided but limited information about observed anisotropy. On the other hand, the widely used cupping test yields a quite satisfactory picture of the sheet's planar anisotropy.

The numerous studies, which have been published in the last thirty years, showed clearly that the development of material design and drawing technology must be preceded by comprehensive analysis of plastic anisotropy relying on the measurement and proper interpretation of the strain ratio. First of all, it is important to determine what type of texture provides the best drawability in the planned plastic working process, and – in turn – which technological process of casting, rolling and heat treatment should be applied to obtain this type of texture in the material.

The differentiation of the rate of deformation in three principal directions in the simple uniaxial tension was considered by Baldwin Jr., Howald and Ross [3.4] as a criterion of plastic anisotropy. The authors were correlating the directionality of the flow properties (relative rates of deformation in three principal directions) with earing tendency of isotropic and anisotropic copper.

In 1947, Lankford, Low and Gensamer [3.5] observed plastic anisotropy in the aluminium alloy. They reported that:

"Some evidence of anisotropy was observed in the materials studied. The rates of contraction in the width and thickness directions in the simple tension test were found to be unequal."

In 1948 Hill [3.6] developing the theory of plastic flow of anisotropic metals suggested the determination of strain increment ratio referred to the principal axes of anisotropy i.e.: $d\varepsilon_w/d\varepsilon_t$. In Hill's opinion, the anisotropy parameters are related to the state of anisotropy immediately preceding the necking occurring in a tensile test. He found that:

"This is effectively the same as in the rolled sheet since the additional anisotropy introduced by the preliminary uniform extension is usually negligible." [3.7]

In 1948, Jackson, Smith and Lankford [3.8] suggested the measuring of anisotropy constants: $K_{yz} = \varepsilon_w/\varepsilon_t$ (for test pieces pulled along the direction of rolling) and $K_{xz} = \varepsilon'_w/\varepsilon_t$ (for test pieces pulled transverse to the direction of rolling), where ε'_w is the strain in the width direction as far as the sample is concerned, but is along the direction of rolling with respect to the original sheet. They claimed that "variation in K_{yz} and K_{xz} for the same material indicate nonhomogeneity". However, in conclusion, Jackson, Smith and Lankford warn that:

"The validity of this procedure is not yet clearly established and additional exploratory experiments designed to clarify the definition of these parameters are in progress. It should be pointed out, however, that, as it will be shown later, even this rather crude method of measuring anisotropy is of considerable value in correlating results from biaxial tests."

From today's perspective, it appears that Hill and Jackson, Smith and Lankford were the closest to the correct definition of plastic anisotropy. Though, it was not until 1949 and 1950 that studies were published showing improved experimental methods leading to the determination of anisotropy coefficient, and providing the possibility of the quantitative measurement of plastic anisotropy in the current analysis of metals and alloys. The crucial contributions were: Krupkowski and Kawiński method [3.9] developed in 1949, and Lankford, Snyder and Bauscher method [3.10] expounded in 1950. The above two studies provide the exclusive reference point in the discussed field of study. However, Krupkowski's study appeared earlier and many authors acknowledge this author's priority. The analysis offered by Lankford et al. [3.10] was accompanied by discussion published by Heyer and Solter [3.11]. They compare the values of the strain ratio of two types of the mild steel calculated from Krupkowski's and Lankford's formulae, finding very similar results. In this way Heyer and Solter prove that Krupkowski's method was known in United States at least since 1950. It is true, though, that Lankford et al. were the pioneers in showing the relation between strain ratio and the drawability of metal sheet. However, systematic investigation of the variation of strain ratio during the tensile test was launched several years later [3.12 – 3.22].

3.2   The measurement of plastic anisotropy coefficient

Normal plastic anisotropy (plastic strain ratio) r_θ in a metal sheet is generally defined as the ratio of two partial strains: $r = \varepsilon_w/\varepsilon_w$ where ε_w and ε_t are the natural plastic strains in the width and thickness directions, respectively, measured in a tensile-tested specimen after unloading. The symbol θ means the angle between straining axis and rolling direction.

Assuming that the volume remains constant after deformation, the axial strain ε is

$$\varepsilon = \varepsilon_w + \varepsilon_t. \tag{3.1}$$

As mentioned above, it is to several authors that we owe the idea of describing the plastic anisotropy through strain ratio. Yet, it was not before the years 1949 – 1950, that studies were published showing experimentally tested methods used in the definition of the anisotropy coefficient and providing the possibility to evaluate quantitatively the plastic anisotropy in current analysis of metals and alloys. The plastic anisotropy coefficient proposed by Krupkowski and Kawiński in 1949 [3.9] is based on the ratio of engineering partial strains:

$$K = \frac{e_w}{e_t} = \frac{(w_0/w)^2 - 1}{(t_0/t)^2 - 1}, \tag{3.2}$$

where w_0 and t_0, as well as w and t, are width and thickness of the cross-sections of the sample before straining and after the deformation ε.

The coefficient K is measured on the sample disrupted in the place where the necking impact is offset by the influence of the head. The authors demonstrated experimentally that the value of K coefficient is independent of the changes in cross-section slenderness within the interval $w_0/t_0 = 1$ to 5.

In 1950, Lankford, Snyder and Baucher [3.10] published a study presenting the method used to define strain ratio which was convergent to Krupkowski's solution:

$$r = \varepsilon_w/\varepsilon_t = \frac{\log(w_0/w)}{\log(t_0/t)}, \tag{3.3}$$

where the values of w and t were found on the sample strained up to the limit of uniform elongation.

In 1962, Jegaden, Voinchet and Rocquet [3.23] defined the coefficient of instantaneous anisotropy with the j coefficient:

$$j = \frac{dw/w}{dt/t} \, . \tag{3.4}$$

In order to compare the results obtained in all three above shown methods, Jegaden, Voinchet and Rocquet [3.23] calculated the values of r, K and j for the studied mild steel and received: $r = 1.40$; $K = 1.44$; $j = 1.72$. Truszkowski conducted similar experiments on aluminium sample at $\theta = 0$, and obtained the values: $r = 0.39$; $K = 0.34$; $j = 0.46$ [3.20].

Aiming at calculation of the anisotropy coefficient as a material characteristic, Truszkowski determined the values r_0, K_0, j_0 by measuring the strain ratio at various deformation levels and extrapolating the obtained functions $r(\varepsilon)$, $K(\varepsilon)$ or $j(\varepsilon)$ to the initial state ($\varepsilon = 0$). Experiments made on a number of *f.c.c.* and *b.c.c.* metals and alloys showed that the value of r_0 (similarly to K_0 and j_0) obtained in result of extrapolation up to zero deformation has a precise physical meaning. He has shown also [3.20] the significant fact that the calculated value of strain ratio was independent of the adopted criterion (Krupkowski, Lankford or Jegaden).

When the formula (3.3) is presented in the form:

$$r = \frac{\log\left(w_0/(w_0 - \Delta w)\right)}{\log\left(t_0/(t_0 - \Delta t)\right)} \tag{3.5}$$

and, according to Jegaden et al. [3.23] we assume that in the interval of small deformations

$$\Delta t = c\Delta w, \tag{3.6}$$

we obtain

$$r = \frac{\log \dfrac{w_0}{w_0 - \Delta w}}{\log \dfrac{t_0}{t_0 - c\Delta w}} = \frac{\varphi(\Delta w)}{\psi(\Delta w)} \, . \tag{3.7}$$

Now we can calculate

$$r_0 = \lim_{\Delta w \to 0} r = \lim_{\Delta w \to 0} \frac{t_0 - c\Delta w}{c w_0 - c\Delta w} = \frac{1}{c} \cdot \frac{t_0}{w_0}. \tag{3.8}$$

In a similar way we can obtain:

$$K_0 = \lim_{\Delta w \to 0} K = \frac{1}{c} \cdot \frac{t_0}{w_0} \tag{3.9}$$

and

$$j_0 = \lim_{\Delta w \to 0} j = \frac{1}{c} \cdot \frac{t_0}{w_0}, \tag{3.10}$$

thus in consequence:

$$r_0 = K_0 = j_0. \tag{3.11}$$

The corroboration of the above result was demonstrated in the mentioned research on aluminium [3.20]. It may be recalled that for $\theta = 0$, $r_a = 0.39$; $K_a = 0.34$; $j_a = 0.46$ (when determined for ε corresponding to the limit of uniform strain). When $\varepsilon = 0$, the obtained values were: $r_0 = 0.31$, $K_0 = 0.30$ and $j_0 = 0.32$.

In the prior research the r_0 value has been determined from the calculated (ε_i, r_i) data by the extrapolation up to the initial state ($\varepsilon = 0$). Recently, a more accurate maximal error method, proposed by Truszkowski and Kloch and described in Chapter 6, is generally used in our research.

While the observed differences between the values of r_a, K_a, and j_a measured at high deformation are significant, it may be assumed that the divergence between r_0, K_0, and j_0 is negligible.

The theoretical analysis and experimental results show also that the value of the studied material's anisotropy coefficient (defining the directionality of its plastic properties) depends on the adopted measure (see formulae 3.3, 3.2, or 3.4). On the other hand, the strain ratio defined for zero strain is independent of the adopted criterion.

The authors of scientific and technological studies generally do not make any difference between r_a value (defined at considerable deformation) and r_0 (the strain ratio corresponding to the initial state, unchanged in effect of deformation occurring in the course of tensile test). In consequence, the question must be asked about the physical meaning of r_a and r_0 values.

As discussed above, the characteristics of plastic anisotropy can be determined through proper interpretation of the tensile test, and through the analysis of relations between crystallographic orientation and anisotropy. When assessing the suitability of material chosen for deep drawing, it is necessary to rely on the plastic anisotropy value measured at critical tensile strain (i.e. at the instability limit); the necking process starts beyond this point, leading to material's fracture. This observation proves the significance of r_a value in the technology of plastic working of materials. On the other hand, when the anisotropy coefficient is determined indirectly (e.g. on the basis of characteristics of crystallographic orientation), it is the r_0 value that is obtained, reflecting the material characteristics unbiased by the error resulting from the instability of the strain ratio in the tensile test.

This phenomenon was discussed also by Michaluk et al. [3.25] in their study on the effect of texture on the r value in heavy gauge tantalum-tungsten plate. These authors found that the r value of Ta 2.5% W plate indicated high strain sensitivity. In effect, the r values obtained on the basis of initial texture coefficients failed to overlap with those measured after tensile deformation.

In 1956, Truszkowski [3.12] proposed the method of evaluation of plastic anisotropy through the analysis of experimentally determined change of the strain ratio in the tensile test. This concept was systematically developed in the following years [3.13, 3.15 – 3.18, 3.20 – 3.22], but it failed to arouse much interest in scientific world. It was not until 1975 that Hsun Hu [3.26, 3.27], who was apparently unaware of earlier studies, launched the investigations of strain dependence of plastic strain ratio. Hsun Hu implemented his experiments on three annealed low-carbon steels, and demonstrated a strong relation between the mean value r_m and the strain in tensile test. In conclusion of several experiments on steels, Hsun Hu proposed the general mechanism of strain ratio variation in relation to strain, depending on the degree of texture in the initial state of the tested material: constant strain ratio for $r = 1$, and diminishing of anisotropy with strain for $r \neq 1$.

Although some of our experiments could be interpreted as corroborating Hu's proposition, it seems that the physical reality is much more complex. When extrapolating the relationship $r = f(\varepsilon)$ back to $\varepsilon = 0$, the value of r_0 is obtained corresponding to anisotropy at initial state. Yet, different types of texture in the initial state (i.e. different texture components and different values of their sharpness) may result in identical r_0 values. In consequence, the impact of fibre texture (caused by straining) may cause differences in the shape of $r = f(\varepsilon)$ function. This fact was fully demonstrated in our experiments.

It is worth mentioning that Hu uses in his analysis the mean value of strain ratio in the sheet plane $r_m = (r_0 + 2r_{45} + r_{90})/4$. (The index 0, 45 and 90 meaning the angle α between the sample axis and the rolling direction). Yet, the application of such average value may be justified only in the case when r changes continuously in relation to α. It can be shown that such relation does not hold in many cases [3.28]. In Chapter 4, we propose the application of standard deviation based on a greater number of r values determined in the sheet plane at different α angles, as a measure of mean value.

3.3   Assessment of anisotropy on the basis of the variation of partial strains

The function $r(\varepsilon)$ describing the change of the coefficient of anisotropy with strain can be considered as key information about plastic anisotropy. This formulation defines the physical interpretation of the adopted measure of anisotropy and its significance in the technology of plastic working of metals and alloys. Therefore, it is recommendable to scrutinize more closely the methods of describing variation of the strain ratio with elongation. In 1976, Kuśnierz [3.29] proposed a method to determine "the adjusted plastic anisotropy coefficient R_k". The proposed method relies on the relation between partial strains ε_w and ε_t and the total, longitudinal elongation ε. Assuming that partial strains are proportional to the total elongation (κ and λ being the proportionality coefficients), Kuśnierz concludes [3.29] that $\kappa + \lambda = 1$, and the corresponding anisotropy coefficient R_k equals:

$$R_k = \frac{\kappa}{\lambda} = \frac{\kappa}{1-\kappa}, \qquad (3.12)$$

However, the assumption of direct proportionality between partial strains and the total elongation ($\varepsilon_w(\varepsilon)$ and $\varepsilon_t(\varepsilon)$), proposed by Kuśnierz and Jasieński [3.29, 3.30] is unacceptable for the entire zone of tensile deformation, since it is clearly incorrect in the range of small deformations. This fact can be observed in Figures 3.1a and 3.2a presented by Kuśnierz [3.29]. Considering the physical interpretation of the functions $\varepsilon_w = f(\varepsilon)$ and $\varepsilon_t = f(\varepsilon)$, the following values must be adopted: $\varepsilon_w(0) = 0$, and $\varepsilon_t(0) = 0$. Thus, it is incorrect to accept the proposition of Kuśnierz and Jasieński [3.29, 3.30] describing the discussed relationship in the following form:

$$\varepsilon_w = A_1 + \kappa\varepsilon, \quad \varepsilon_t = B_1 + \lambda\varepsilon, \qquad (3.13)$$

when admitting the values of A_1 and B_1 different from zero.

The opinion formulated by the above authors who claim that „the constants A_1 and B_1 are the effect of, among other factors, the unavoidable experimental scatter" is physically meaningless.

Yet, Kuśnierz's experimental results, presented in Figs 3.1a and 3.2a, can be interpreted differently when three distinct mechanisms of plastic

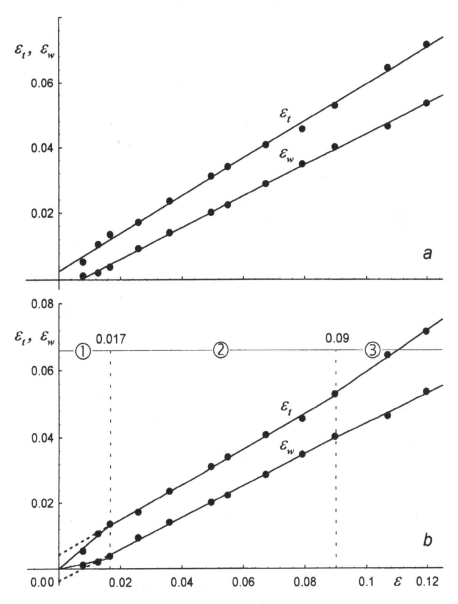

Figure 3.1. a) ε_w vs ε and ε_t vs ε relationships of the copper Cu-0 sample described by a straightlinear function (J.Kuśnierz [3.29]). b) The same experimental data [3.29] described by three different zones (W.Truszkowski and J.Kloch [3.40])

deformation are accepted, occurring in three elongation ranges (Figs 3.1b and 3.2b).

Then, in the case of a copper test piece Cu-0 (Fig.3.1b) the borders between ranges are defined at $\varepsilon_{1/2} = 0.017$ and $\varepsilon_{2/3} = 0.090$, while for the copper test piece Cu-90 (Fig.3.2b): $\varepsilon_{1/2} = 0.043$ and $\varepsilon_{2/3} = 0.093$. For the test

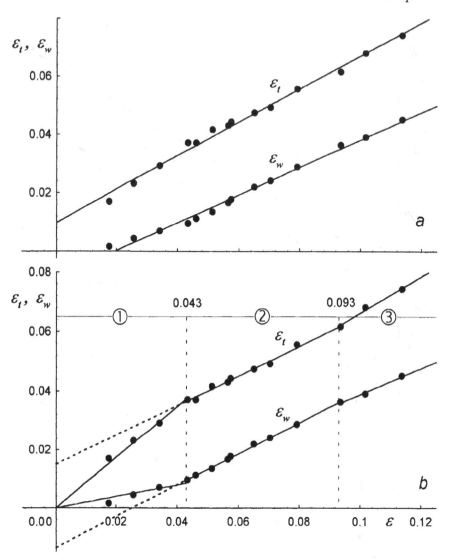

Figure 3.2. a) ε_w vs ε and ε_t vs ε relationships of the copper Cu-90 sample described by a single straightlinear function (J.Kuśnierz [3.29]). b) The same experimental data [3.29] described by three different zones (W.Truszkowski and J.Kloch [3.40])

piece Cu-0, the partial strain change accompanying the elongation in the first deformation zone is defined by the functions

$$\varepsilon_w = 0.197\,\varepsilon; \quad \varepsilon_t = 0.803\,\varepsilon, \tag{3.14}$$

while the anisotropy coefficient

$$r_0 = \frac{\kappa_1}{\lambda_1} = 0.25 . \tag{3.15}$$

For the test piece Cu-90 we obtain:

$$\kappa_1 = 0.192, \quad \lambda_1 = 0.808 \tag{3.16}$$

and

$$r_0 = 0.24. \tag{3.17}$$

3.4 Plastic anisotropy at inhomogeneous deformation

Kuśnierz's experimental results concerning partial strain changes versus total elongation function [3.29] fail to confirm the author's hypothesis that the shortening (partial strain) occurs in direct proportionality to elongation in the entire elongation zone. (In this case, in diagrams 3.1a and 3.2a, the functions $\varepsilon_w = f(\varepsilon)$ and $\varepsilon_t = f(\varepsilon)$ would be described by a single linear relationship). Yet, the only physically meaningful assumption is that at zero elongation partial strains are equal to zero. This fact suggests a different interpretation of Kuśnierz's experimental results [3.29], based on the concept of the macroinhomogeneous deformation. The corresponding diagram consists of three segments of a straight line with different slopes and discontinuities at the limits of particular ranges (see Figs 3.1b and 3.2b). This behaviour, which has been observed by many authors [3.31 – 3.35], can be explained by the discontinuous change of the underlying physical mechanism.

A similar change of the deformation mechanism was observed at tensile test [3.19] of a cold rolled sample of brass sheet (Fig. 3.3). Both relationships $\varepsilon_w(\varepsilon)$ and $\varepsilon_t(\varepsilon)$ can be described by two segments of a straight line. The distinct border between the ranges appears at $\varepsilon_{1/2} = 0.085$. Although, the measurement results in the first range are not numerous, the implement of the linear interpretation of both functions, seems justified at the assumption of $\varepsilon_w(0) = \varepsilon_t(0) = 0$.

The analysis of the deformation of a titanium test piece indicates the change of the mechanism in the process of progressive straining at room temperature [3.36]. The deformation starts from slip (in different crystallographic systems) and leads to twinning and to dynamic strain ageing. The shift from one mechanism to the other is accompanied by discontinuity in the stress – strain relationship, as well as in the r vs ε function. It is more meaningful to analyse the deformation ranges by plotting $\log(d\sigma/d\varepsilon)$ against $\log\varepsilon$ (σ stands for true stress). The diagram in Fig. 3.4 shows the results obtained at room temperature of the test piece cut out in the rolling direction

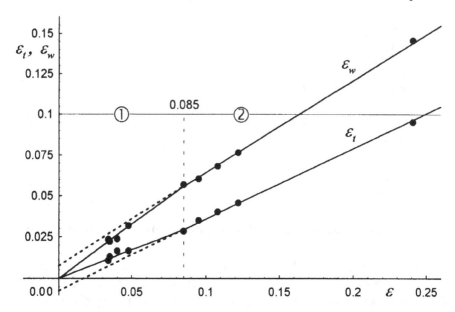

Figure 3.3. ε_w vs ε and ε_t vs ε relationships of the cold rolled CuZn19 brass. Rolling reduction $z = 40.5\%$, orientation of the sample axis $\alpha = 90°$ to the rolling direction. Experimental data from [3.19]

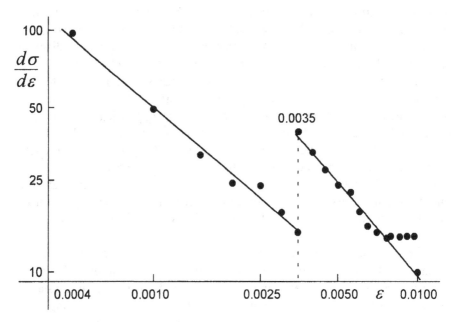

Figure 3.4. $d\sigma/d\varepsilon$ vs ε relationship of the annealed titanium; σ is true stress. Orientation of the sample axis in the sheet $\alpha = 0°$ to the rolling direction. Experimental data from [3.36]

of technical purity titanium sheet. In the zone of low deformation (up to $\varepsilon = 0.01$, i.e. 1% of elongation), the relation very clearly shows two ranges of straightlinear function. The discontinuity at $\varepsilon = 0.0035$ reflects the abrupt shift from slip-type deformation to twinning. This stage is shown in $\varepsilon_w(\varepsilon)$ and $\varepsilon_t(\varepsilon)$ diagrams by the change of slope of the straight line (Fig. 3.5).

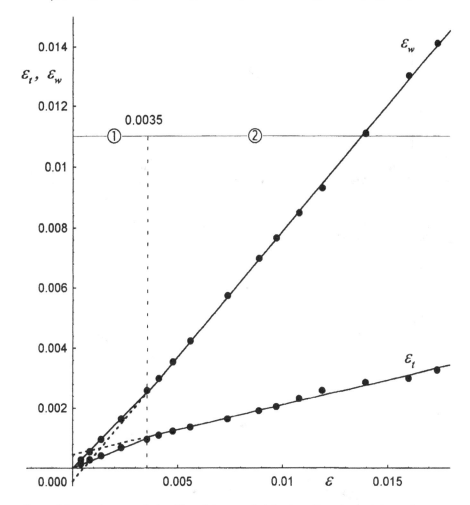

Figure 3.5. ε_w and ε_t vs ε relationships of the annealed titanium. Experimental data as in Fig. 3.4

Even the CuZn16 brass [110] single crystal with high level of crystallographic orientation [3.37] (where the half-width of the neutron rocking curve was equal to 50') indicated two straining ranges. Both of them could be described by linear functions of significantly different slopes. It is difficult to identify precisely the border between ranges as observed at very low elongation ($\varepsilon_{1/2} = 0.00085$). Nevertheless, the interpolation of a significant

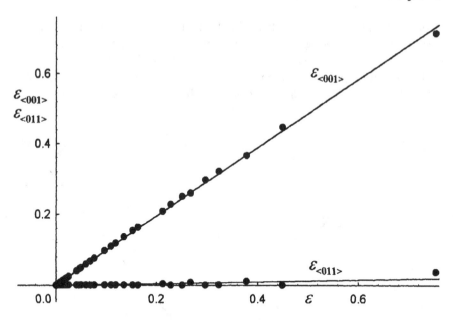

Figure 3.6. ε_w vs ε and ε_l vs ε straightlinear relationship of the CuZn16 [110] single crystal. Experimental data from [3.37]

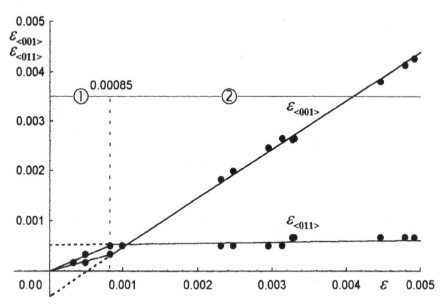

Figure 3.7. The diagram from Fig. 3.6 restricted to the zone $\varepsilon \in [0, 0.005]$

number of points (at very low experimental scatter), and the extrapolation of the so-obtained straight line from the second to the first range yields the finite values $A_1 = 0.0005$ and $B_1 = -0.0005$ (3.14). The diagram of $\varepsilon_w(\varepsilon)$ and

$\varepsilon_i(\varepsilon)$ functions in the entire elongation zone of a strained crystal is shown in Fig. 3.6. The border between ranges 1 and 2 is graphically displayed by the same relationship limited to the zone between zero and $\varepsilon = 0.01$ (Fig.3.7).

3.5 Discussion of results

The characteristics of plastic anisotropy can be defined on the basis of tensile test results $(\varepsilon_{wi}, \varepsilon_i)$ and $(\varepsilon_{ti}, \varepsilon_i)$ data in two ways.

In the former approach (1) ε_w and ε_t values are found for any step of the tensile test in order to define the $r(\varepsilon)$ relationship. In this way, the impact of elongation on strain ratio is established. It has been shown [3.38] that the $r(\varepsilon)$ function can be described by a hyperbola. The finding of appropriate curve parameters makes it possible to calculate r_0 precisely. Thus, the plastic anisotropy of the studied material is provided with well defined physical meaning.

In the latter approach (2), the relation of partial strains to the total elongation is established. Yet, $\varepsilon_w = f(\varepsilon)$ and $\varepsilon_t = f(\varepsilon)$ do not define plastic anisotropy. In order to obtain this characteristics, it is necessary to relate two functions: $\varepsilon_w = \kappa\varepsilon$ and $\varepsilon_t = \lambda\varepsilon$, where the κ/λ coefficient represents the mean value of anisotropy in the entire elongation zone. The κ/λ coefficient is identical with r_0 only in the case of fully homogeneous deformation in tensile test. However, this is observed but in exceptional cases. It is noteworthy that even in single crystals of brasses, the authors [3.37] identified two ranges of the stress – strain (and consequently in r – strain) relationship. The macroinhomogeneity of deformation at the tensile test is revealed in the discontinuity of the $r(\varepsilon)$ function, and the appearance of two or more deformation ranges.

The function describing the variation of the strain ratio in the course of elongation (approach 1) exhibits high scatter of experimental points in the zone of small ε values. It is important to observe that this range is essential for the determination of r_0. Yet, the calculation of $r(\varepsilon)$ relationship with the use of physically corroborated hyperbolic function [3.38] must be based on experimental data in the zone of high deformation, in which the scatter of measured r values is much smaller. Relying on this procedure, we found in [3.37] the following values in the case of four <011> brass single crystals: $r_0 = 0.5; 0.5; 0.48; 0.55$ (the average value being $\bar{r}_0 = 0.508$). In spite of the significant scatter around the $r(\varepsilon)$ function at the initial deformation (from $r = 0.2$ to $r = 1.0$), the results confirm the theoretically found value for f.c.c. metals $r_{0\,[110]} = 0.5$ [3.39, 3.40].

It could be expected that the latter approach (2) is useful in the case of perfectly macrohomogeneous deformation at the tensile test. Then, $\varepsilon_w(\varepsilon)$ and $\varepsilon_t(\varepsilon)$ are described by linear functions in the entire range of uniform elongation: $\varepsilon_w = \kappa\varepsilon$, $\varepsilon_t = \lambda\varepsilon$, and $\kappa/\lambda = r_0$. However, when the deformation is macroinhomogeneous, and the first zone is relatively small, the finding of r_0 is difficult and technically inaccurate, also due to high experimental scatter.

3.6   Final remarks and conclusions

The former method of interpretation of anisotropy findings (i.e. the analysis of $r(\varepsilon)$ relation described by the hyperbolic function) is expressed in: (i) the finding of the r value at a high strain (which is useful for calculation of technological processes, e.g. of deep drawing), or (ii) calculation of r_0, which is the inherent material property. It must be emphasized that correct data are obtained even in the case of macroinhomogeneous deformation at the tensile test.

The latter method (i.e. the plotting of $\varepsilon_w(\varepsilon)$ and $\varepsilon_t(\varepsilon)$ functions) can yield full characterization of plastic anisotropy only in the case of macrohomogeneous deformation of a strained test piece. The analysis of these functions at the onset of straining of all materials tested by the present authors (polycrystalline metals as well as single crystals) has shown at least two ranges. This is the evidence of macroinhomogeneity of deformation. It is worth mentioning that the anisotropy coefficient evaluated in this way describes only a mean value. Yet, the straightlinear course of $\varepsilon_w(\varepsilon)$ and $\varepsilon_t(\varepsilon)$ relations (in limited ranges) corroborates the hyperbolic model of $r(\varepsilon)$ function [3.38].

In order to compare the accuracy of r_0 calculated through both mentioned methods, we have applied the results obtained in the case of CuZn16 [110] single crystal: (a) theoretically found $r_{0\,[110]} = 0.5$ [3.38, 3.39]; (b) $r_{0\,[110]}$ obtained from the analysis of the experimental $r(\varepsilon)$ function: $r_0 = 0.55$ (though the mean value of four brass [110] single crystals was $r_0 = 0.508$ [3.37, 3.40]; (c) calculated according to the latter method (i.e. by plotting $\varepsilon_w(\varepsilon)$ and $\varepsilon_t(\varepsilon)$ experimental points in the first zone of elongation): $r_0 = \kappa/\lambda = 0.645$ [3.40].

The above findings lead to the conclusion that the analysis of $r(\varepsilon)$ function (i.e. the former method) is recommendable in view of the inhomogeneity of deformation in real materials.

3.7   References

3.1. Ch.S.Barrett and T.B.Massalski, *Structure of Metals. Crystallographic methods, principles and data, Third edition.*, *McGraw-Hill Book Company*, New York, 540, (1966).

3.2. H.W.Brownsdon, *Discussion at Birmingham*, J. Inst. Met., **60**, 178 (1937).

3.3. J.D.Jevons, *Discussion at Birmingham*, J. Inst. Met., **60**, 174 (1937).

3.4. W.M.Baldwin Jr., T.S.Howald and A.W.Ross, *Relative Triaxial Deformation Rates*, Trans. Am. Inst. Mining Metall. Eng., **166**, 86 (1946).

3.5. W.T.Lankford, J.R.Low and M.Gensamer, *The Plastic Flow of Aluminium Alloy Sheet under Combined Loads*, Trans. Am. Inst. Mining Metall. Eng., **171**, 574 (1947).

3.6. R.Hill, *A theory of the yielding and plastic flow of anisotropic metals*, Proc. Roy. Soc., Series A, **193**, 218 (1948).

3.7. R.Hill, *The Mathematical Theory of Plasticity*, Oxford 1971 (first published in 1950).

3.8. L.R.Jackson, K.F.Smith and W.T.Lankford, *Plastic Flow in Anisotropic Sheet Steel*, Metals Technology. Am. Inst. Mining Metall. Eng. August, 1948, T.P.2440.

3.9. A.Krupkowski and S.Kawiński, *The Phenomenon of Anisotropy in Annealed Polycrystalline Metals*, J. Inst. Metals, **75**, 869 (1949).

3.10. W.T.Lankford, S.C.Snyder and J.A.Bauscher, *New Criteria for Predicting the Press Performance of Deep Drawing Sheets*, Trans. Am. Soc. for Metals, **42**, 1197 (1950).

3.11. R.H.Heyer and R.L.Solter, *Written Discussion (of paper 3.10)*, Trans. Am. Soc. for Metals, **42**, 1226 (1950).

3.12. W.Truszkowski, *Zagadnienie anizotropii zgniecionych metali polikrystalicznych*, Arch. Hutn., **1**, 171 (1956).

3.13. W.Truszkowski et Z.Bojarski, *Sur l'anisotropie de l'acier inoxydable 18/8 laminé à froid*, Mém. Sci. Rev. Métallurg., **59**, 112 (1962).

3.14. G.Pomey et M.Grumbach, *Quelques corrélations entre les coefficients d'anisotropie et d'écrouissage et les essais d'emboutissage*, Rev. de Mét., **61**, 885 (1964).

3.15. W.Truszkowski, *On the Plastic Anisotropy of Metals Defined by the Strain Ratio*, Bull. Acad. Pol. Sci., sér. techn., **15**, 717 (1967).

3.16. W.Truszkowski, *Experimental Contribution to the Method of Measuring the Strain Ratio of Metals*, Bull. Acad. Pol. Sci., sér. techn., **15**, 805 (1967).

3.17. W.Truszkowski and J.Jarominek, *Plastic Anisotropy of Cold Rolled Copper*, Arch. Hutn., **14**, 309 (1969).

3.18. W.Truszkowski et J.Król, *Sur l'évaluation quantitative de l'anisotropie des métaux*, C.R.Acad. Sci. Paris., **269**, 807 (1969).

3.19. W.Truszkowski, J. Dutkiewicz and J. Szpunar, *Evolution de la texture et de l'anisotropie lors du laminage du laiton*, Mém. Sci. Rev. Métallurg., **67**, 355 (1970).

3.20. W.Truszkowski, *Sur le sens physique du rapport des allongements obtenu par la méthode d'extrapolation*, La Metalurgia Italiana, **24**, 489 (1969).

3.21. W.Truszkowski et J.Król, *Tendance á l'anisotropie des métaux cubiques à faces centrées laminés à froid*, Mém. Sci. Rev. Métallurg, **67**, 201 (1970).

3.22. W.Truszkowski et J.Jarominek, *Essai de synthèse des recherches sur l'anisotropie plastique*, Mém. Sci. Rev. Métallurg., **70**, 433 (1973).

3.23. G.Jegaden, J.Voinchet et P.Rocquet, *Contribution à l'étude de la déformation plastique des tôles*, Mém. Sci. Rev. Métallurg., **59**, 273 (1962).

3.24. W.Truszkowski, *On the Quantitative Evaluation of Plastic Anisotropy in Sheet Metals*, Proc. 8th Biennial Congress of IDDRG, Gothenburg, 48 (1974).

3.25. Ch.Michaluk, J.Bingert and C.S.Choi, *The Effects of Texture and Strain on the R - Value of Heavy Gauge Tantalum Plate*, Mem. Sci. Forum, **157 – 162**, 1653 (1994), Textures and Materials, Proc. ICOTOM-10, Clausthal, 1993.

3.26. Hsun Hu, *The Strain Dependence of Plastic Strain Ratio (r_m value) of Deep Drawing Steel Sheets Determined by Simple Tension Test*, Met. Trans., **6A**, 945 (1975).

3.27. Hsun Hu, *Effect of Plastic Strain on the r - Value of Textured Steel Sheet*, Met. Trans., **6A**, 2307 (1975).

3.28. W.Truszkowski, *Influence of Strain on the Plastic Strain Ratio in Cubic Metals*, Met. Trans., **7A**, 327 (1976).

3.29. J.Kuśnierz, *Właściwa wartość współczynnika anizotropii*, Rudy Metale, **21**, 15 (1976).

3.30. J.Kuśnierz et Z.Jasieński, *Courbe de traction et valeur propre du coefficient d'anisotropie des tôles d'aluminium, du cuivre et de laiton*, Mém. Sci. Rev. Métallurg., **73**, 485 (1976).

3.31. A.M.Garde, R.E.Reed-Hill, *Dual Analysis of Longitudinal and Transverse Zirconium Tensile Stress – Strain Data*, Special Technical Publication 551, Am. Soc. Test. Mater., 75 (1974).

3.32. W.Truszkowski, A.Łatkowski and A.Dziadoń, *Stress – Strain Behaviour and Microstructure of Polycrystalline Alpha-Titanium*, Proc. 2nd Risø. Intern. Symp. on Metallurgy and Mat. Science, 383 (1981).

3.33. R.E.Reed-Hill, W.R.Cribb and S.Monteiro, *An Empirical Analysis of Titanium Stress – Strain Curves*, Met. Trans., **4**, 1011 (1973).

3.34. R.E.Reed-Hill, W.R.Cribb and S.Monteiro, *Concerning the Analysis of Tensile Stress – Strain Data Using* $\log d\sigma/d\varepsilon_p$ *Versus* $\log \sigma$ *Diagrams*, Met. Trans., **4**, 2665 (1973).

3.35. S.Krishnamurthy, K.W.Qian and R.E.Reed-Hill, *Effects of Deformation Twinning on the Stress – Strain Curves of Low Stacking Fault Energy Face-Centered Cubic Alloys*, Practical Applications of Quantitative Metallography, ASTM STP 839, Philadelphia, 41 (1984).

3.36. W.Truszkowski and S.Wierzbiński, Institute for Metal Research PAS, (Unpublished).

3.37. W.Truszkowski, A.Modrzejewski and J.Baczyński, *Variation of the Strain Ratio in Tensile Tested [110] Brass Single Crystals*, Bull. Pol. Ac.: Techn., **37**, 471 (1989) .

3.38. W.Truszkowski and J.Kloch, *The Variation of Strain Ratio at the Tensile Test Described by a Hyperbolic Function*, Textures and Microstructures, **26 – 27**, 531 (1996).

3.39. A.Krupkowski, *Anizotropia mono– i polikrystalicznego metalu o strukturze Al*, Arch.Hutn., **2**, 9 (1957).

3.40. W.Truszkowski and J.Kloch, *New Aspects of Plastic Anisotropy in Metals*, Bull. Pol. Ac.: Techn., **46**, 299 (1998).

Chapter 4

The quantitative evaluation of plastic anisotropy

4.1   Introduction

As it has been mentioned earlier, in annealed pure metals and single phase alloys the only factor ultimately determining the plastic anisotropy is the preferred crystallographic orientation. Hence, since the texture can be described in terms of quantity, also a fully quantitative description of the anisotropy of the plastic properties should be possible. This hypothesis created the possibility of experimental verification of the physical meaning of the coefficient A_p, proposed already in the year 1969 by Truszkowski and Król [4.1, 4.2] as a measure of plastic anisotropy:

$$A_p = (1 - |D_{0m}|)\sqrt{\frac{\sum_{i=1}^{n}(D_{0i} - D_{0m})^2}{n-1}} \quad , \tag{4.1}$$

where:

$$D_0 = \frac{r_0 - 1}{r_0 + 1} . \tag{4.2}$$

D_{0m} is the mean value of the function $D_0(\theta)$, where θ is the angle between the rolling direction (RD) of the metal sheet and the axis of the sample.

A measure of plastic anisotropy may be also the mean value of the strain ratio in the examined metal sheet. When applying the author's own concept of the limiting value of r_0 according to (3.9), $\bar{r}_0$ can be determined [4.3] from the formula:

$$\bar{r}_0 = \frac{1}{2\pi} \int_0^{2\pi} r_0(\theta) d\theta \; . \tag{4.3}$$

Confrontation of plastic anisotropy with crystallographic texture requires a quantitative representation of each of these phenomena by means of one parameter. As a measure of texture there can be used the index J after Sturcken and Croach [4.4], defined on the basis of the orientation distribution function $f(g)$ or the coefficient T_c according to Truszkowski and Król [4.1, 4.2], calculated for *f.c.c.* metals from the pole figure $\{111\}$:

$$J = \oint (f(g) - 1)^2 dg \; . \tag{4.4}$$

For a random orientation the texture index $J = 0$, for a perfect single crystal $J = \infty$.

The coefficient T_c is the variance of pole density on a $\{111\}$ pole figure:

$$T_c = \sum_{i=1}^n P_i (1 - I_i)^2 \; , \tag{4.5}$$

where P_i is a fraction of an equal-area $\{111\}$ pole figure with the mean pole density I_i. Here, for a random orientation: $T_c = 0$, and for a perfect single crystal $T_c = \infty$.

The A_p value refers to a state unchanged under the influence of deformation in the tensile test itself; hence, in case of annealed single phase alloys this value may be compared with texture characteristics (e.g. T_c), since the method of X-ray or neutron diffraction does not induce changes in the material which might cause a change in the crystallographic orientation.

4.2   Plastic anisotropy in polycrystalline *f.c.c.* metals

In order to modify the experimental concept of a quantitative evaluation of plastic anisotropy there have been carried out investigations on polycrystalline copper close to the random orientation, cold rolled from the annealed state up to a high deformation degree [4.1]; for the particular deformation degrees there have been determined the values of A_p and T_c after the formulae (4.1) and (4.5). The investigation results are graphically presented in Fig. 4.1, where z is a relative reduction of the cross-section.

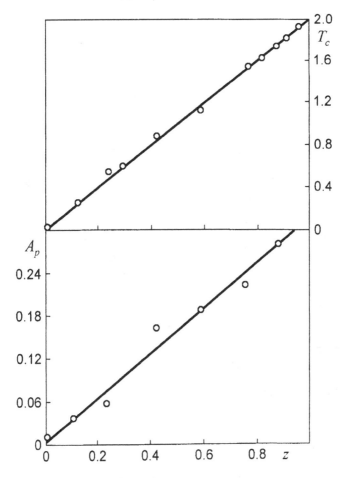

Figure 4.1. Effect of rolling reduction on T_c and A_p coefficients in copper

To the initial state (which is close to the isotropy condition) there correspond the values of A_p and T_c, close to zero, in each case the function $A_p(z)$ and $T_c(z)$ being linear. On the graphs the straight lines determine the course of the dependence for copper rolled from the isotropic state: $A_p = 0.312z + 0.0056$ and $T_c = 2.00z + 0.0047$. It can be observed that the same character of the function (a linear function), describing the change in the characteristics of plastic anisotropy and texture in a single phase material, such as pure copper, justifies the physical meaning of the coefficient A_p. This conclusion is supported by the results obtained in the investigations of nickel [4.1] – Fig. 4.2, notwithstanding a somewhat greater distance from the random orientation observed in the annealed state: $A_p = 0.209z + 0.0068$ and $T_c = 0.924z + 0.0867$.

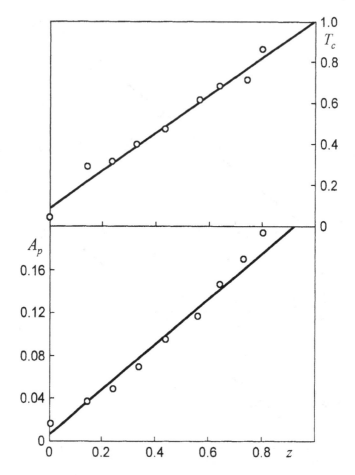

Figure 4.2. Effect of rolling reduction on T_c and A_p coefficients in nickel

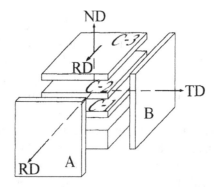

Figure 4.3. Schematic presentation of the positions of A, B and C planes in the examined metal plate

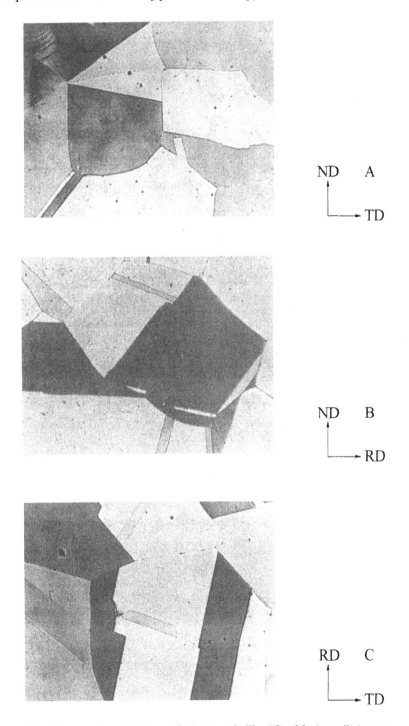

Figure 4.4a. Microstructure in A, B and C planes (as in Fig. 4.3) of the hot rolled copper (×150)

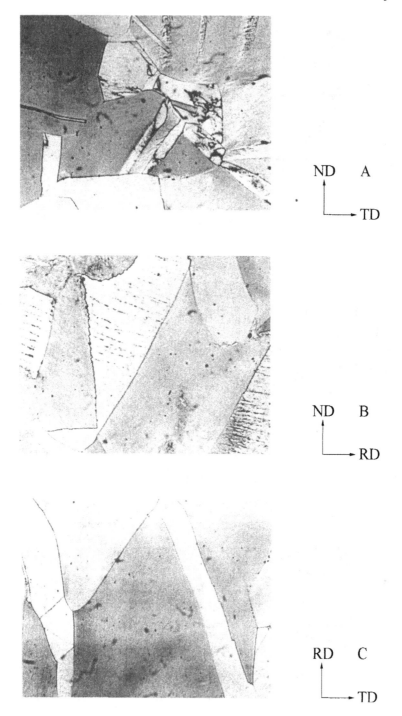

Figure 4.4b. Microstructure in A, B and C planes (as in Fig. 4.3) of the cold rolled copper (x150)

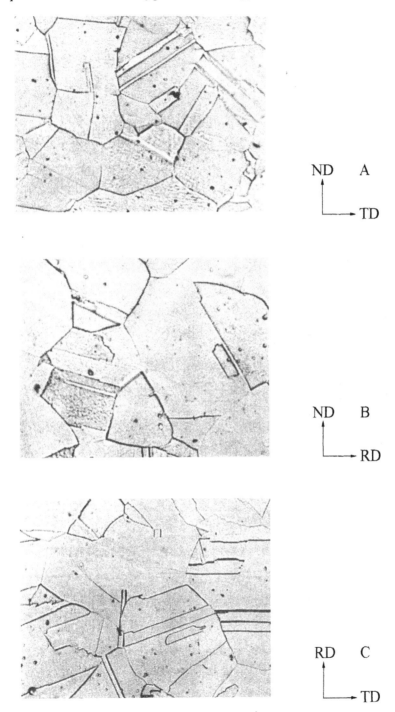

Figure 4.5a. Microstructure in A, B and C planes (as in Fig. 4.3) of the hot rolled nickel (x150)

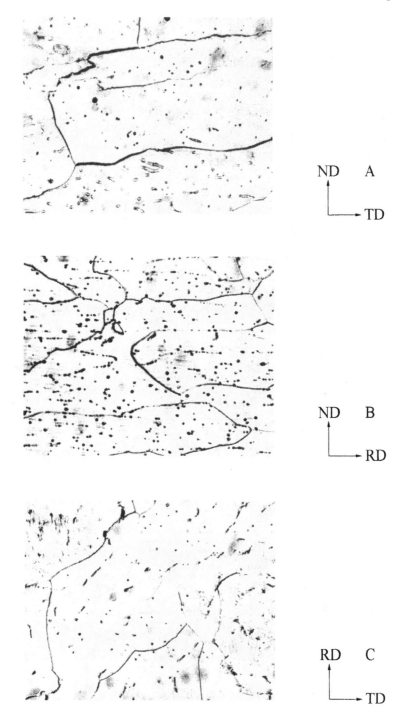

Figure 4.5b. Microstructure in A, B and C planes (as in Fig. 4.3) of the hot rolled aluminium (x150)

Thus, the coefficient A_p has a strictly defined physical meaning representing the measure of the plastic anisotropy of the materials. The pole figure of the coefficient r_0, which is usually determined experimentally for a plane parallel to the rolling plane, fully describes the plastic anisotropy of the material. Thus, if in order to calculate the value of A_p we assume as a basis the characteristics of the changes of the coefficient r_0 in another plane selected from the sample system (e.g. perpendicular to the rolling direction or to the transverse direction), we will obtain, using the formula (4.1) the same value of A_p (obviously assuming the macrohomogeneity of the material).

This thesis has found support in the experiments on the change of the coefficient r_0 in three planes perpendicular to the basic directions in hot rolled copper, nickel and aluminium plates [4.5]: ND – normal direction to the rolling plane C, RD – rolling direction normal to plane A, and TD – transverse direction (normal to the plane B).

The position of the planes A, B and C in the examined thick sheet of metal is shown in Fig. 4.3, and the microstructures presenting the form of crystals in the particular sections of the materials – in Figs 4.4 and 4.5. From appropriately prepared plates, cut out from thick sheets of copper, nickel and aluminium, there were prepared samples of circular cross-section, the axis of which deviated by the angle 0°, 22.5°, 45°, 67.5° and 90° from ND (in the case of plates cut out from the sheet parallel to the planes A and B) or from RD (in case of plates C). The samples were subjected to the tensile test and the functions $r(\varepsilon)$ determined. Their extrapolation to $\varepsilon = 0$ enabled to deter-

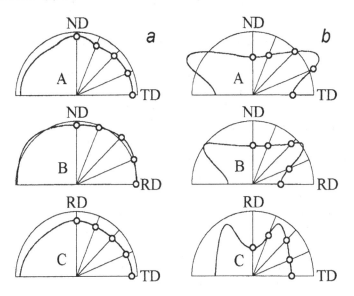

Figure 4.6. r_0 pole figures of hot rolled copper (*a*) and cold rolled copper (*b*)

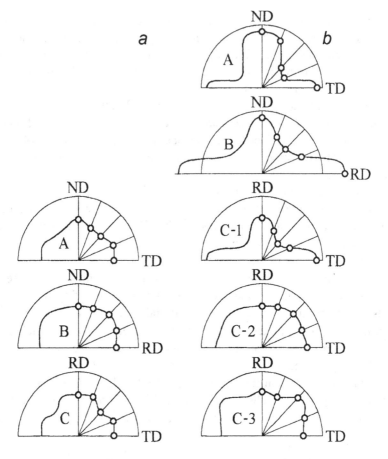

Figure 4.7. r_0 pole figures of hot rolled nickel (*a*) and hot rolled aluminium (*b*)

mine the value of r_0. Since only in aluminium a fairly distinct macroinhomogeneity of the material, manifested by the lamellar structure of the examined
thick sheet, could be observed, the measurements were carried out on three
plates parallel to the rolling plane: C-1 from the middle of a thick sheet, C-2
from the intermediate layer, and C-3 from the surface layer (Fig. 4.3).

   The results collected on the pole figures of the coefficient r_0 for the three
examined metals illustrate the distribution of the anisotropy characteristics in
three planes A, B and C (Figures 4.6 and 4.7). These data allowed to
calculate, using formula (4.1), the value of A_p on the basis of experimental
data collected from these three planes. For comparison, there have been
carried out calculations of the coefficient $(A_p)_a$ under conditions in which as
a measure of anisotropy there has been accepted the value r_a of this coefficient, which is attained by the material at the limit of uniform elongation.
The respective values are listed in Table 4.1. The results show very good

Table 4.1.

Material	Plane	A_p	$(A_p)_a$
Hot rolled copper	A	0.0171	0.0068
	B	0.0184	0.0217
	C	0.0186	0.0187
Cold rolled copper	A	0.0992	0.0966
	B	0.0995	0.0982
	C	0.0997	0.0933
Hot rolled nickel	A	0.0240	0.0077
	B	0.0235	0.0136
	C	0.0236	0.0144
Hot rolled aluminium	A	0.1367	0.1421
	B	0.1373	0.1886
	C–1	0.1115	0.0931
	C–2	0.0170	0.0226
	C–3	0.0413	0.0447

agreement between the A_p values for copper and nickel for all examined plates A, B and C; the difference between the maximal and the minimal value in the case of copper does not exceed 9%, 2% in the case of nickel and 0.5% in the plates A and B of aluminium. It can be assumed that these divergences are within the limits of experimental error. On the other hand, accepting in the calculations the value r_a causes that the error increases to 209%, 100% and 107%, respectively. The results contribute to a justification of the physical meaning of r_0 determined by the method of extrapolation to the initial state. As it has been demonstrated earlier, the difference between the values A_p and $(A_p)_a$ depends on the type and the degree of texture in the sample subjected to tension.

4.3 Plastic anisotropy in *f.c.c.* single crystals

4.3.1 Anisotropy coefficient r_0 in perfect [100], [110] and [111] single crystals

The method of calculation of the coefficient $r_0 = d\rho_x/d\rho_y$, based on a geometrical model of octahedral slip in *f.c.c.* metals when taking into account the principle of minimum deformation work, was elaborated by Krupkowski [4.6]. For the orientation [100] and [111]: $r_{0[100]} = 1$ and $r_{0[111]} = 1$, and for the single crystal [110]: $r_{[110]} = 0.5$. This result has been supported experimentally by Krupkowski and Cichocki [4.7] in a tensile test of aluminium [110] single crystal; the straining caused a drop of the value of the strain ratio from 0.53 to 0.22, and the function $r(\varepsilon)$ extrapolated linearly up to $\varepsilon = 0$ gave $r_0 \cong 0.57$. The small number of measurements and the great experimental scatter around the function $r(\varepsilon)$ – discussed in Chapter 5 –

justify the occurrence of some error in the extrapolated value, nevertheless the value of r_0 is close to 0.5.

However, as much as the plastic strain ratio values for the orientations [100] and [111], calculated by the method of Krupkowski, are in agreement with the quantities determined on the basis of a strictly geometrical model, in case of orientation [110] neglecting the principle of minimal work leads to a great error, yielding the value $r_{0\,[110]}$ equal to zero.

Such an incorrect result has been obtained by Roberts [4.8] when calculating the anisotropy coefficient on the basis of the geometrical model of octahedral slip. He writes:

"In face-centered cubic metals, it is well known that recrystallization can, under certain conditions produce «cube texture», a recrystallization texture in which cube planes lie parallel to the surface of the sheet with a cube direction parallel to the rolling direction. The texture is so sharp that the sheet may be regarded as a pseudo single crystal. The values of the "R" coefficient obtained from copper sheet consisting of almost 100 per cent cube texture are given in Table 4.2 and are in good agreement with the predicted single crystal values."

Systematic investigations of the change in the strain ratio on tensile tested real single crystals of aluminium, copper and silver with orientations changing along the edge of the basic stereographic triangle from [100] to [110] (orientations [100], [110], [510], [310], [210], [320] and [110]) have been carried out by Truszkowski, Gryziecki and Jarominek [4.9, 4.10, 4.11]. Thus, e.g. tensile testing of crystal [110] showed a fast drop of r from the value 0.5 to 0.08 in aluminium, to 0.05 in copper and to 0.03 in silver. In all cases extrapolation of the experimentally determined $r(\varepsilon)$ function in the range of small deformations gave – as it has been shown in Chapter 5 – the values close to $r_{0\,[110]} = 0.5$.

Table 4.2. (part) After W. T. Roberts [4.8]

Direction of the sample	
Parallel to [100] Parallel to [010]	Parallel to [110]
Anticipated coefficient r	
1.0	0.0
Coefficient r measured in a sheet of almost 100% cubic texture	
1.02	0.07

Finally, the application of the author's own method, described in Chapters 5 and 6, of calculating r_0 from experimental data, based on the assumption of a hyperbolic function and the weights determined from the magnitude of experimental scatter, has resulted in the calculation of the value $r_0 = 0.48 \div 0.55$ for [110] brass single crystals with the zinc contents of 9.1%, 10%, 16% and 30% [4.12, 4.13]. It is to be noticed that the mean value $r_{0[110]} = 0.508$ is very close to the calculated value. These results are reported in Chapter 7.

Thus, it is to be concluded that the application of the method of calculating the anisotropy coefficient, proposed by Roberts, which in this case is not physically justified, and the incorrect interpolation of the experimental results disregarding the anisotropy changes in the tensile test, gave a wrong evaluation of the anisotropy coefficient for [110] *f.c.c.* single crystals. On the other hand, the agreement between the anticipated value $r = 1.0$ for the orientation [100] and the measured $r = 1.02$ in the same pseudo–monocrystalline copper sheet (Table 4.2) is fully satisfactory. As it has been demonstrated earlier, the range of stability for the orientation [100] is considerably greater and it is easier to obtain a sample for which $r(\varepsilon) = \text{const} = 1.0$.

A critical analysis of Robert's results shows once more that in a material of unstable (in a tensile test) crystallographic orientation (i.e. in most *f.c.c.* single crystals), which are the object of our experiments, it is not possible to determine, in a direct way, the material feature such as the anisotropy coefficient r_0. This problem is important in so far as many authors show the tendency to evaluate the strain ratio basing on a strictly geometrical model, which – as it has been demonstrated above – may lead to great errors.

4.3.2 Plastic anisotropy in real single crystals

In perfect single crystals the coefficients describing the degree of anisotropy and "texture" (A_p, J and T_c) calculated according to formulae (4.1), (4.4) and (4.5) – have infinite value, whereas in real single crystals a certain dispersion around the nominal orientation gives in effect their finite values. As a result a real single crystal may be considered as a polycrystalline aggregate with very sharp, one-component texture.

Aluminium single crystals. A quantitative description of the "texture" of a real crystal by means of X-ray technique using the method applied to polycrystalline materials is not very effective on account of its unusual sharpness. Accordingly, the method of extrapolation was applied [4.14]: a single crystal is deformed (e.g. by rolling) and the texture characteristics $T_c = f(z)$ or $T = f(z)$ for increasing deformation degree z (z is the reduction in area) is determined. The function $T_c(z)$ (or $J(z)$), found experimentally enables to calculate the looked-for value by way of its extrapolation to the initial state ($z = 0$). In this way the value $T_0 = 178$ has been found for the analyzed aluminium [100] single crystal.

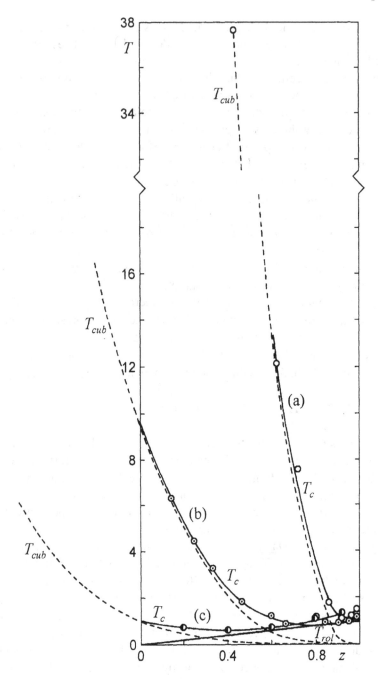

Figure 4.8. $T_c(z)$ (full line), $T_{cub}(z)$ (dashed line) and $T_{rol}(z)$ (full straight line) relationships for: [100] aluminium single crystal (**a**), aluminium with sharp cube texture (**b**), aluminium with weak cube texture (**c**) and aluminium with close-to-random orientation (T_{rol})

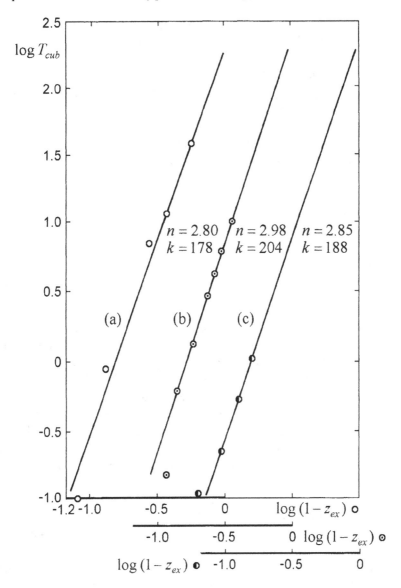

Figure 4.9. T_{cub} vs $(1 - z_{ex})$ relationship. Denotations as in Fig. 4.8

The study was carried out on aluminium samples in three states of intensity of the cubic texture [4.15]: on polycrystalline aluminium at $T_0 = 1.04$
and $T_0 = 9.97$, and on the above mentioned single crystal in which the value
T_0, determined by the above described method, was equal to $T_0 = 178$. The
results in the form of the relation $T_c(z)$ were referred to the earlier calculated
[4.1, 4.2] course of $T_c(z)$ for aluminium with the random orientation ($T_0 = 0$).
The linear dependence for the rolling of aluminium with the random orienta-

tion ($T_{rol}(z)$) [4.2], enabled, adopting some additional auxiliary assumptions, to estimate the value $T_0 = 178$ for the examined single crystal (Fig. 4.9).

The following experimentally supported assumptions were of help here:

1. description of the dependence $T_{cub}(z)$ by the power function in the entire deformation range z; T_{cub} is a quantitative estimation of the texture component with cubic orientation.

2. assumption $T_c(z) = T_{cub}(z) + T_{rol}(z)$, where T_{rol} is a component of rolling texture, typical for f.c.c. metals with high stacking fault energy.

The value T_0 for aluminium single crystal can be determined from the relation

$$T_{cub} = k(1 - z_{ex})^n,$$ (4.6)

where

$$z_{ex} = z_0 + (1 - z_0)z.$$ (4.7)

Good agreement between the course of the experiment and equation (4.6) for all three states of intensity of the crystallographic orientation in the examined aluminium samples is shown by the graph in Fig. 4.9.

Brass single crystals. Single crystals of α brass of the [110] orientation were grown by the modified Bridgman technique. Since zinc is a volatile component of brass, crystals can be grown from the melt only under inert gas pressure. This can be achieved by the use of a high-pressure steel autoclave or sealed steel ampoules. A device with wire-wound resistance heater and sealed stainless steel ampoule was used. The alloy was melted in a graphite crucible.

Three different ways of the preparation of the starting materials from the elements followed by directional solidification of the melt have been tested. In the first case, zinc and copper were melted in a graphite crucible and crystals were grown from the melt without the use of steel ampoule. The loss of material during solidification reached 50 pct of initial weight; the higher the zinc content, the greater was the material deficiency. In the second set of experiments pieces of zinc and copper of a required composition were melted in a crucible sealed in a steel ampoule. Crystals were grown at a loss of initial materials not greater than 2 pct and increasing with the increase of furnace temperature. This was eliminated by immersing small pieces of zinc into liquid copper in argon atmosphere and fast solidification of the alloy. Crystals were then grown in sealed steel ampoule. Melting of the component elements in an arc furnace resulted in a considerable evaporation of zinc.

Crystals of α brass of 9.1, 10, 16 and 30 wt pct zinc content were grown at a lowering rate of the crucible 10 mm/h. The diameter of the crystals was

22 and 44 mm and the length – 30 to 50 mm. Since they were grown in a sealed ampoule, direct observation of the melt and also the measurement of temperature in the vicinity of the crucible were not possible. Therefore, the growth conditions were chosen experimentally in a set of processes. The most important thing was not to preheat the melted alloy; preheating of the melt resulted in intensive penetration of zinc into the steel wall of the ampoule and in increased evaporation of zinc.

The chemical analysis has shown that crystals are homogeneous in composition. Specimens cut out from the first and last parts to freeze show difference in zinc content below 0.1 pct.

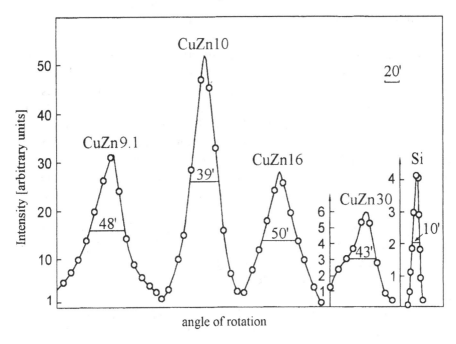

Figure 4.10. Neutron rocking curves measured in reflection for the four α brass single crystals at the 022 reflection and for the silicon crystal at the 111 reflection [4.13]

The mosaic structure of brass crystals was investigated by neutron diffraction studies. The crystal under investigation was set at the centre of the rotating table of a spectrometer and examined by means of the neutron beam reflected from the crystal monochromator and passing through a 30' Soller collimator located behind it. Experiments were performed in reflection at the 220 plan in Bragg position. The results were obtained in the form of rocking curves (Fig. 4.10). Full width at half maximum of the rocking curve, called the half-width (ρ), gave information about the mosaic structure of the crystal. The value of the mosaic spread (χ) of the crystal can roughly be estimated from the expression given by

$$\chi = \sqrt{\rho^2 - M^2} .$$ (4.8)

M being the half-width of the experimental resolution function calculated from the rocking curve for a single crystal with a very low mosaic spread, e.g. silicon.

Neutron diffraction experiments have shown that the grown brass crystals reveal an increased spread of the mosaic structure compared with pure copper crystals. The half-widths of the neutron rocking curves for the four studied crystals varied from 39 to 50 minutes of arc (CuZn9.1 – 48', CuZn10 – 39', CuZn16 – 50', CuZn30 – 43'). For the same geometry of the assembly the half-width for 111 silicon crystal was 10', and for copper crystals – about 20'.

Four tensile test specimens of the [110] orientation have been prepared with the accuracy estimated at 0.5° (the angle $\alpha \le 0.5°$ between [110] and the axis of the sample in the {001} plane). Round section test pieces of $d_0 = 6$ mm diameter and about 20 mm gauge length were used. After mechanical working they were chemically etched and electrolytically polished. The tensile test was carried out at room temperature with constant crosshead motion 0.2 mm min^{-1}, interrupted periodically for the measurement of the contour of the cross section of the specimen at three sections AA, BB (in the middle of the gauge length) and CC. The measuring accuracy was $2 \cdot 10^{-3}$ mm.

The direct method of calculating the strain ratio has been used which is the most adequate for a precise establishing of the $r(\varepsilon)$ relationship [4.16].

Nickel single crystals. When studying the imperfection of the crystallographic orientation of *f.c.c.* single crystals, the investigations were carried out on samples with symmetrical orientation with respect to the tensile axis [100], [111] and [110]. In copper and brass single crystals produced by the Bridgman method the degree of perfection better than $\rho \approx 20'$ could not be obtained, whereas in nickel single crystals, produced using the method of Czochralski, modified by Kyropoulos, the ρ value was established at $\rho \approx 10'$.

Nickel single crystals were obtained by pulling of the crystals from liquid metal [4.17]. To obtain the assumed orientation there have been used oriented nuclei, placed in a water cooled holder. The holder with the nucleus was dropped till it got into contact with the melted metal, whose temperature was kept 50° up to 10° above the melting temperature. Increase of temperature of the melted nickel caused partial melting of the nucleus before pulling out of the single crystal was begun.

Nickel was melted in argon atmosphere in a graphite crucible, placed under a quartz bowl, with a heating coil wound on its outside. The holder with the nucleus and a thermocouple were placed inside the bowl. In order to avoid non-uniform growth of the crystal caused by asymmetric distribution of temperature, the holder with the nucleus was rotated at a rate of 15 up to 20 revolutions per minute. Considering the thermal resistance of the single crystal, increasing with time, its cooling was intensified by reducing the temperature of the liquid metal. The length of a single crystal reached 120 mm with the diameter of 9 to 12 mm.

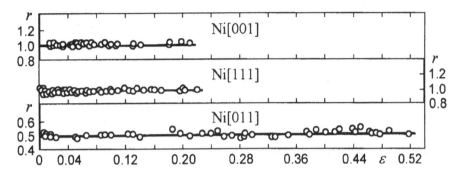

Figure 4.11. $r(\varepsilon)$ relation for tensile tested nickel single crystals grown by the Czochralski method [4.17]

Fig. 4.11 shows that due to high degree of ordering of the crystallographic orientation in nickel single crystals obtained by the Czochralski method, the $r(\varepsilon)$ function was constant in the case of all three orientations: $r_{[100]} = 1.0$, $r_{[110]} = 0.5$ and $r_{[111]} = 1.0$.

4.4   Final remarks

Plastic anisotropy as an intrinsic property of the material should be calculated on the basis of the r_0 value – the value of strain ratio at zero strain. The method of its determination is described in Chapter 6. On the other hand, the strain ratio calculated at large deformation (e.g. at the limit of uniform elongation) is an important technological property describing the directionality of plastic properties in heavy deformed material, and is useful for the calculation of the process of deep drawing of metal sheets.

In single crystals, the strain ratio can be theoretically calculated when starting from the crystallographic orientation [u v w]; the thus determined value represents the intrinsic property of a perfect crystal and is equal to its r_0 value. In real single crystals, however, structure and orientation faults have a distinct impact on the r value, and therefore it is justified to adopt the notion of the "texture of a crystal". The influence of different factors, defining the deviation of a real single crystal from the perfect structure on "texture in single crystal" and their mechanical behaviour and instability of orientation at the uniaxial strain is the subject of Chapter 9.

4.5 References

4.1. W.Truszkowski et J.Król, *Sur l'évaluation quantitative de l'anisotropie des métaux*, C.R. Acad. Sci. Paris, **269**, 807 (1969).

4.2. W.Truszkowski et J.Król, *Tendance à l'anisotropie des métaux cubiques à faces centrées laminés à froid*, Mém. Sci. Rev. Métallurg., **67**, 201 (1970).

4.3. P.Parnière et G.Pomey, *Relations entre l'anisotropie cristallographique et l'anisotropie mécanique. Cas des tôles minces d'acier extra doux pour emboutissage*, Mécanique, Matériaux, Electricité, **XI**, 1 (1974).

4.4. E.F.Sturcken and J.W.Croach, *Predicting Physical Properties in Oriented Metals*, Trans. Met. Soc. AIME, **227**, 934 (1963).

4.5. W.Truszkowski et J.Jarominek, *Essai de synthèse des recherches sur l'anisotropie plastique*, Mém. Sci. Rev. Métallurg., **70**, 433 (1973).

4.6. A.Krupkowski, *Anizotropia mono— i polikrystalicznego metalu o strukturze A1*, Arch. Hutn., **2**, 9 (1957).

4.7. A.Krupkowski and S.Cichocki, *Shape Deformation of the Single Crystals of Aluminium in the Tensile Test*, Polish Academy of Sciences, Commission of Metallurgy and Foundry, **3**, (1969).

4.8. W.T.Roberts, *Crystallographic Aspects of Directionality in Sheet*, Sheet Met. Ind., **39**, 855 (1962).

4.9. W.Truszkowski, J.Gryziecki and J.Jarominek, *Assessment of the Strain Ratio in the Cube Plane of f.c.c. Metals*, Bull. Ac. Pol. Sci., sér. techn, **24**, 209 (1976).

4.10. W.Truszkowski, J.Gryziecki and J.Jarominek, *Variation of Strain Ratio in Cube Plane of Copper*, Metals Technology, **6**, 439 (1979).

4.11. W.Truszkowski, J.Gryziecki and J.Jarominek, *Variation of Plastic Strain Ratio in the {001} Crystallographic Plane of Silver*, Bull. Pol. Ac.: Techn, **31**, 31 (1983).

4.12. W.Truszkowski, A.Modrzejewski and J.Baczyński, *Variation of the Strain Ratio in Tensile Tested <011> Brass Single Crystals*, Bull. Pol. Ac.: Techn., **37**, 471 (1989).

4.13. W.Truszkowski and A.Modrzejewski, *Influence of Stacking Fault Energy on Instability of Crystallographic Orientation in Tensile Tested Brass Single Crystals*, Arch. of Metallurgy, **35**, 219 (1990).

4.14. W.Truszkowski, *The Impact of Texture in Single Crystals of FCC Metals on Mechanical Behavior and Instability of Orientation*, Proc. 8th Intern. Conf. on Textures of Materials (ICOTOM – 8). Ed J.S.Kallend, G.Gottstein, Santa Fe, 1988, 537.

4.15. W.Truszkowski, *Sur les états extrêmes d'orientation crystallographique dans les métaux c.f.c.*, Rev. int. Htes Temp. et Réfract., **14**, 65 (1977).

4.16. W.Truszkowski, S.Wierzbiński and A.Modrzejewski, *Influence of Mosaic Structure on Instability of the Strain Ratio in Deformed Copper Single Crystals*, Bull. Acad. Pol. Sci., sér. techn., **30**, 367 (1982).

4.17. W.Truszkowski, S.Wierzbiński, A.Modrzejewski, J.Baczyński, G.S.Burhanov, I.V.Burov, and O.D.Čistjakov, *Influence of Deviation from <001>, <011> and <111> Orientations on the Variation of Strain Ratio in Deformed Nickel Single Crystals*, Arch. of Metallurgy, **32**, 165 (1987).

Chapter 5

The variation of strain ratio at the tensile test described by a hyperbolic function

5.1 The two ranges $r(\varepsilon)$ function

Plastic deformation at the tensile test of the polycrystalline metal sample with the random crystallographic orientation induces the formation of texture that reproduces the symmetry of the deformation process: the sample axis is the axis of symmetry of the forming texture. It has been shown in experiments that in case of random (or close-to-random) orientation $r(\varepsilon)$ is a linear function, the r_0 value being close to 1; this can be illustrated on example of recrystallized nickel in the close-to-isotropic state, and annealed copper for

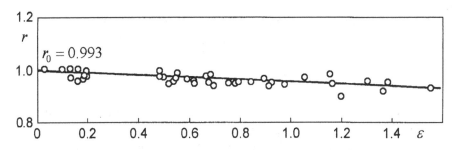

Figure 5.1. $r - \varepsilon$ diagram for the annealed nickel. Experimental data from [5.1]

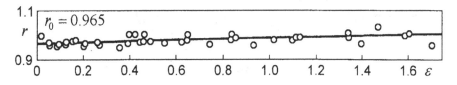

Figure 5.2. $r - \varepsilon$ diagram for the annealed copper. Experimental data from [5.2]

which the X-ray diffraction revealed almost random orientation. The respective functions are: $r = 0.993 - 0.046\varepsilon$ (for nickel sample – Fig 5.1) and $r = 0.965 + 0.0365\varepsilon$ (for copper sample – Fig. 5.2) [5.1, 5.2].

A similar behaviour is observed in single crystals with highly ordered, symmetric crystallographic orientation. Figures 5.3 and 5.4 present $r - \varepsilon$ relations of the [100] copper crystal grown by the Bridgman method [5.3] and [110] nickel crystal produced using the method of Czochralski [5.4]. For both samples the strain ratio vs. strain relationship is described by the straight line: $r(\varepsilon) = 1$, in the case of the [100] copper single crystal [5.3], and $r(\varepsilon) = 0.5$, for the [110] nickel crystal [5.4]. The parameters of all samples have been calculated using the procedure described in Chapter 6.

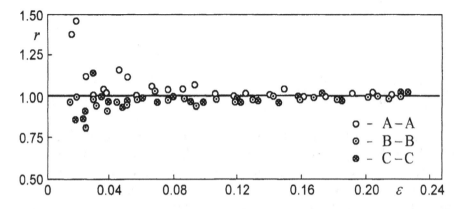

Figure 5.3. $r - \varepsilon$ diagram for the [100] copper single crystal. Experimental data from [5.3]

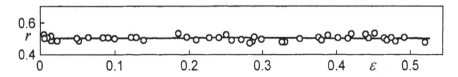

Figure 5.4. $r - \varepsilon$ diagram for the [110] nickel single crystal. Experimental data from [5.4]

If, however, the sample of the polycrystalline metal with sharp texture and the coefficient r_0, differing much from the value 1 (e.g. the hot-rolled nickel which $r(\varepsilon)$ function is shown in Figure 5.5) is subdued to tensile test, the $r - \varepsilon$ diagram exhibits two ranges: ① increase in r to attain the value $r \sim 1$, and ② slow diminishing of the strain ratio r. It is characteristic that in both ranges $r(\varepsilon)$ function can be described by the straightlinear function; its slope in the first range is n_1, while in the second n_2, the transition from n_1 to n_2 occurs in a continuous way. The $r(\varepsilon)$ function of the polycrystalline nickel, described above, illustrates well the variation of the strain ratio with the elongation in the tensile test, independently of whether in the first range one observes the increase or decrease of the *r* value.

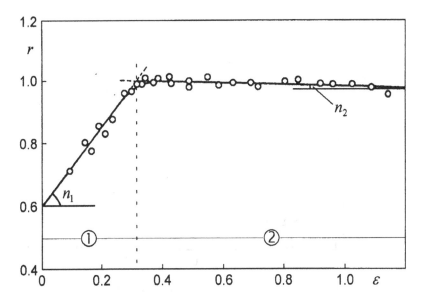

Figure 5.5. Unstable behaviour at the tensile test of the hot rolled polycrystalline nickel. Experimental data from [5.5]

Analogically, two ranges in the $r - \varepsilon$ curve can be observed in the process of straining of a single crystal in which crystal orientation deviates from the symmetrical one. Fig. 5.6 presents the $r - \varepsilon$ relationship for the CuZn10-4° [110] brass single crystal (the [110] direction deviated from the sample axis by 4° in the [110] plane): the rapid, close-to-linear, drop of *r* with ε up to about $\varepsilon = 0.004$ followed by slow, equally linear diminishing of *r* values within range $\varepsilon > 0.02$, and continuous variation of the slope of the $r(\varepsilon)$ function in a transition range $0.004 < \varepsilon < 0.02$. Such a variation of the strain ratio with strain has suggested to the authors [5.6] the description of the $r(\varepsilon)$ function by a hyperbola.

The equation of the hyperbola defined for all $\varepsilon \in R$ has the form:

$$r = a_1 \varepsilon + a_2 + a_3 \sqrt{\varepsilon^2 + a_4 \varepsilon + a_5} \qquad (5.1)$$

if only:

$$a_4^2 - 4a_5 \leq 0. \qquad (5.2)$$

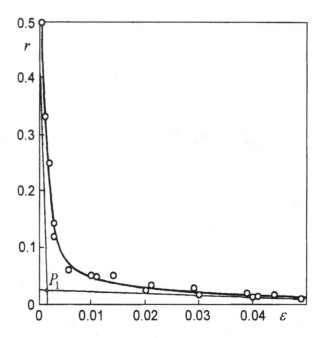

Figure 5.6. Unstable behaviour at the tensile tested 4° [110] CuZn10 brass single crystal. Experimental data from [5.6]

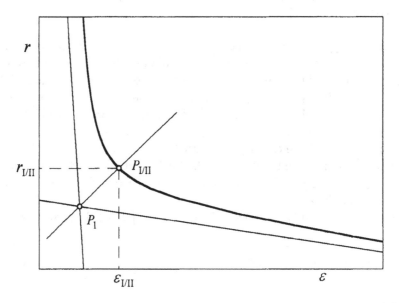

Figure 5.7. Determination of the boundary ($P_{I/II}$) between the first (I) and the second (II) deformation ranges in the $r(\varepsilon)$ function. The $P_1 - P_{I/II}$ line is the bisector of the angle between asymptotes

The equations of asymptotes are:

$$r = (a_1 + a_3)\,\varepsilon + a_2 + 0.5\,a_3 a_4$$
$$r = (a_1 - a_3)\,\varepsilon + a_2 - 0.5\,a_3 a_4 \tag{5.3}$$

and the point of their intersection is:

$$P_1(\varepsilon_1, r_1) = (-0.5\,a_4,\ a_2 - 0.5\,a_1 a_4). \tag{5.4}$$

The equation (5.1) describes each case of the $r(\varepsilon)$ function met in practice: from the arbitrary position of asymptotes on the $\varepsilon - r$ plane to the straight line (when the a_3 parameter is equal to zero). The value $r_0 = r(0)$ is equal to $r_0 = a_2 + a_3\sqrt{a_5}$.

The boundary between the ranges in the $r(\varepsilon)$ function ($P_{I/II}$) can be determined as is shown in the diagram in Fig. 5.7:

$$x_{I/II} = -\frac{a_4}{2} + \frac{a_3}{2}\sqrt{\frac{4a_5 - a_4^2}{(A - a_1)^2 - a_3^2}}, \tag{5.5}$$

$$y_{I/II} = A\,\frac{a_3}{2}\sqrt{\frac{4a_5 - a_4^2}{(A - a_1)^2 - a_3^2}} + a_2 - \frac{a_1 a_4}{2}, \tag{5.6}$$

where

$$A = -\frac{\sqrt{(a_1 - a_3)^2 + 1} + \sqrt{(a_1 + a_3)^2 + 1}}{(a_1 + a_3)\sqrt{(a_1 - a_3)^2 + 1} + (a_1 - a_3)\sqrt{(a_1 + a_3)^2 + 1}}. \tag{5.7}$$

5.2 Experimental verification

5.2.1 Single crystals

Single crystal of 10 wt pct Zn brass with the orientation 4° [110] has been tensile tested up to elongation $\varepsilon = 0.05$ (Fig 5.6). The parameters of the equation (5.1) calculated for the tested brass with the least square method are: $a_1 = -102.09$, $a_2 = 0.2209$, $a_3 = -101.73$, $a_4 = -3.8\ 10^{-3}$, $a_5 = 7.5\ 10^{-6}$.

Consequently, the asymptotes are: $r = -0.3577\varepsilon + 0.0269$, $r = -203.88\varepsilon + 0.4151$ and their intersection point $P_1(\varepsilon_1, r_1) = (0.002, 0.026)$.

It can be seen that the hyperbola (5.1) proposed for the description of results plotted in the $r(\varepsilon)$ relationship fits well the experimental points

(Fig. 5.6). For $\varepsilon = 0$ the value $r_0 = r(0)$ is equal to 0.5 which is in a perfect agreement with the theoretically calculated value $r_{0\,[110]} = 0.05$. [5.7].

Equally good fitting by the hyperbolic equation (5.1) of experimental data from the tensile test of the [110] copper single crystal is observed in Fig. 5.8.

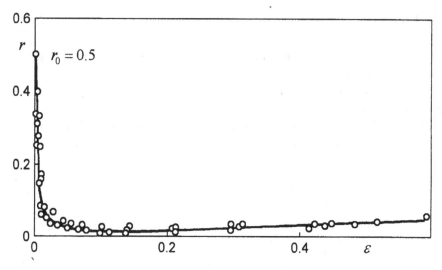

Figure 5.8. $r(\varepsilon)$ relationship of the [110] copper single crystal. Experimental data from [5.8]

The high efficiency of the proposed method of joining the hyperbolic description with the maximal error procedure was shown in the research of tensile testing of four brass single crystals [9.19]. The chemical composition of crystals varied from 9.1 to 30.0 wt % Zn and their crystallographic orientation was [110]. In spite of large experimental scatter around the $r(\varepsilon)$ function the average value is $\bar{r}_{0\,[011]} = 0.508$ which is in good accordance with the theoretical value (Figs 9.13 – 9.16).

5.2.2   Polycrystals

Several test pieces of polycrystalline *f.c.c.* metals and alloys have been strained up to the instability limit and the hyperbolic fitting function was calculated using the above described method. The examples presented in Chapter 6 – (Figs 6.6 – 6.12) evidence the correct approximation of the $r - \varepsilon$ data by the hyperbolic function. All $r(\varepsilon)$ functions have been calculated using the earlier proposed method of fitting function based on the concept of maximal error. The results constitute a convincing verification of the proposed method.

5.3 References

5.1. W.Truszkowski und A.Łatkowski, *Über die Anisotropie des verformten vielkristallinen Nickels*, Freiberger Forschungshefte, **B 123**, 61 (1967).

5.2. W.Truszkowski and J.Jarominek, *Plastic Anisotropy of Cold Rolled Copper*, Arch. Hutn., **14**, 309 (1969).

5.3. W.Truszkowski i S.Wierzbiński, *Izmenene koefficienta plastičeskoj anizotropii monokristallov medi s orientacjej blizkoj [001] pri rastjazenii*, Fizika Metallov i Metallovedenie, **56**, 1195 (1983).

5.4. W.Truszkowski, S.Wierzbiński, A.Modrzejewski, J.Baczyński, G.S.Burhanov, I.V.Burov and O.D.Čistjakov, *Influence of Deviation from <001>, <011> and <111> Orientations on the Variation of Strain Ratio in Deformed Nickel Single Crystals*, Arch. Metall., **32**, 165 (1987).

5.5. W.Truszkowski et J.Jarominek, *Essai de synthèse des recherches sur l'anisotropie plastique*, Mém. Sci. Rev. Métallurg., **70**, 433 (1973).

5.6. W.Truszkowski and J.Kloch, *Instability of Strain Ratio in Single Crystals Defined by the r(ε) Function*, Bull. Pol. Ac.: Techn., **37**, 467 (1989).

5.7. W.Truszkowski, *Quantitative Aspects of the Relation between Texture and Plastic Anisotropy*, Proc. Intern. Conf. on Textures of Materials (ICOTOM 7), Noordwijkerhout, 723 (1984).

5.8. W.Truszkowski, J.Gryziecki and J.Jarominek, *Variation of Strain Ratio in Cube Plane of Copper*, Metals Technology, **6**, 439 (1979).

Chapter 6

Determination of parameters of the $r(\varepsilon)$ function from experimental data

by J.Kloch and W.Truszkowski

6.1 Characteristics of experimental scatter along the $r(\varepsilon)$ function

Strain ratio describing plastic anisotropy is usually determined in the tensile test by measuring the variation of the dimensions of the cross-section of the strained test piece. As in the case of any experiment the results are burdened with error its effect being the scatter of the calculated values. When assuming the value of maximum error c (when measuring the dimensions of varying cross-section) the maximum error in the strain ratio r at any step of the tensile test can be evaluated. In result the zone of maximum scatter around the $r(\varepsilon)$ function (r being strain ratio, ε – elongation) can be determined.

Let us assume that the deformation at the uniaxial tensile test is macro-homogeneous; this is conditioned by the stability of experimental error at the tensile test $(c(\varepsilon) = \text{const.})$. Theoretical considerations and experimental evidence [6.1 – 6.3] show that error in the r value and consequently the scatter around the $r(\varepsilon)$ function is large in the zone of small elongations, rapidly diminishes with the elongation and then slowly rises in the range of extreme strains (which cannot be attained in the practice of the tensile test) when the dimension of the cross-section approaches the numerical value of the c – parameter. As a total elongation of a polycrystalline metal (as well as in single crystals) rarely exceeds $\varepsilon = 2$, only the first zone of the diminishing scatter around the $r(\varepsilon)$ function should be considered. However, having in view the small influence of experimental error on the ε value (the rectangle of error is very slender) this effect was not considered.

When the deformation at the tensile test is macroinhomogeneous and the experimental error rises with the strain (i.e. c rises with strain) the scatter of r values is considerable even in the zone of medium deformations. Such behaviour can be observed in coarse grained and/or in strongly textured metals; in this case the regular cross-section of the test piece (circular, elliptic, square...) changes into irregular one and the determination of w and t dimensions becomes more and more difficult (or even impossible). Similar phenomenon has been observed at the tensile test of the [110] single crystal of aluminium [6.4]; the high stacking fault energy and the low strain hardening exponent hinder to activate sufficient number of slip systems at the onset of deformation.

6.2   Determination of the fitting function based on maximal errors

6.2.1   Introduction

Let us consider the following typical problem concerning the interpretation of the experimental data. For a given sequence of points (x_i, y_i) on a plane, being a result of direct or indirect measurements, we seek for a function $y = f(x)$ that would provide the best approximation of the points (x_i, y_i), and also would enable us to perform the correct extrapolation in the zone where measurements were not or could not be made. In the case when the values of errors as well as the form of the function under consideration are known, the method of the least squares is usually applied to solve the problem.

If, however, additional difficulties appear such as:
i) large scatter of the experimental points, whose magnitude is dependent on the variable x;
ii) remarkable asymmetry of scatter of the experimental points with respect to the fitting function $f(x)$; i.e. $f(x_i)$ is not the centre of the interval of maximal error $[y_i, y_i]$ (Fig. 6.1);
iii) inability to determine the values of some errors resulting from the applied method of measurement;
iv) lack of theoretical premises as to the form of the function $f(x)$; however, such a situation rarely takes place because in many cases the form of the function is known;
then no general method of determination of $f(x)$ exists.

This problem has arisen in connection with the difficulties in the determination of the $r(\varepsilon)$ function (r being the strain ratio, ε – logarithmic strain) from the experimental results. When assuming the magnitude of error in measuring r and ε, the formulae have been deduced for the functions describing the limits of the zone of maximal error around $r(\varepsilon)$. The problem is of considerable importance as the scatter of r values is very large at small strains and $r_0 = r(0)$ is often experimentally determined using the back extrapolation method.

Here the procedure is presented, which in some cases enables us to determine $f(x)$ in spite of all additional difficulties (i – iv) appearing simultaneously. The range of its application is defined by the assumptions formulated in Sect. 6.2.2. This method makes possible not only the determination of $f(x)$ but also the calculation of the values of some experimental errors.

The essential feature of the presented method is that it is based not on mean but on maximal errors and the related zone of maximal error (Fig. 6.1). This has both advantages and disadvantages. For example, the advantage of the method is that the asymmetry of scatter can be fully taken into account in the process of approximation; on the other hand it may happen that an experimental point remarkably departing from the others can considerably affect the final result, which is of course the disadvantage.

The method consists of three stages. The purpose of the first stage is to get a general idea of the shape of experimental scatter and, consequently, to draw two functions $f^-(x)$ and $f^+(x)$ being the rough approximation of the lower and upper limitation (Fig. 6.1) of the zone of maximal error along the function $f(x)$ to be found. In the second stage, taking advantage of $f^-(x)$ and $f^+(x)$, we obtain two auxiliary sequences of points $(x_i, \bar{y}_i)$ and $(x_i, \bar{c}_i)$ which facilitate the determination of the form of both functions $f(x)$ and $c(x)$ that gives the evaluation of the unknown errors. In the third stage, we fit the values of the parameters of the functions determined in the second stage in such a way that zone of maximal error would contain all experimental points, being at the same time the narrowest among those which possess this property. A more precise description of the particular stages will be given in Sect. 6.2.3.

6.2.2   Notations and assumptions

Suppose that from direct measurements we obtain $u_i = (u_{i1}, u_{i2}, \ldots, u_{ik})$ and $v_i = (v_{i1}, v_{i2})$, $i = 1, 2, \ldots, n$; $k \geq 1$. The related maximal errors are denoted respectively: $\delta u_i = (\delta u_{i1}, \delta u_{i2}, \ldots, \delta u_{ik})$ and $\delta v_i = (\delta v_{i1}, \delta v_{i2})$. Suppose further that we have got two real functions $x(u, v) = x(u_1, \ldots, u_k, v_1, v_2)$ and $y(u, v) = y(u_1, \ldots, u_k, v_1, v_2)$ defined in an open region $D \subset R^{k+2}$.

Assumptions:
a) the values of the maximal errors δu_i are known;
b) the values of the maximal errors δv_i are unknown but satisfy the following equalities

$$\delta v_{i1} = \delta v_{i2}, \qquad i = 1, 2, \ldots , n; \tag{6.1}$$

c) the sample size n is greater than $10 \, (k + 2)$;
d) the functions $x(u, v)$ and $y(u, v)$ are of class C^1 in the open region D;
e) Jacobian $J = D(x, y)/D(v_1, v_2)$ and all partial derivatives of the functions $x(u, v)$ and $y(u, v)$ are different from zero in the region D.

Evaluation of the sample size n, given in the *assumption* c) is only the rough one, since what is satisfactory for the simpler cases when we already know the form of the function $f(x)$, being only interested in the determination of its parameters, may appear unsatisfactory when applied to the more complicated cases with a rather large and asymmetrical scatter of experimental points, additionally if the form of the function $f(x)$ is unknown.

Under the *assumption* e) the functions $x(u, v)$ and $y(u, v)$ are monotonic with respect to each variable, separately; hence they reach their extrema at the vertices of any cuboid contained in D.

From the *assumptions* d) and e) it follows that for every fixed u the function $v \rightarrow (x, y) = (x(u, v), y(u, v))$ is one-to-one; therefore, there exists its inverse function

$$(x, y) \rightarrow v = v(u; x, y) \tag{6.2}$$

The function $v(u; x, y)$ will be helpful in the determination of the zone of maximal error along the function $f(x)$.

Let us denote

$$\Delta v_i = \sum_{j}^{2} [v_{ij} - c, v_{ij} + c] = [v_{i1} - c, v_{i1} + c] \times [v_{i2} - c, v_{i2} + c] \subset R^2, c > 0$$

$$\Delta u_i = \sum_{j=1}^{k} [u_{ij} - \delta u_{ij}, u_{ij} + \delta u_{ij}] \subset R^k, \qquad \delta u_{ij} > 0 \tag{6.3}$$

For such $(u, \delta u, v, c)$ that $\Delta u \times \Delta v \subset D$ we can calculate:

$$y^-(u, \delta u, v, c) = y^-(X) = \min\{y(s, t) : (s, t) \in \Delta u \times \Delta v \subset D\}$$

$$y^+(u, \delta u, v, c) = y^+(X) = \max\{y(s, t) : (s, t) \in \Delta u \times \Delta v \subset D\} \tag{6.4}$$

An interval $[y^-(X), y^+(X)]$ will be called the interval of maximal error with respect to y at the point $(x(u, v), y(u, v))$.

Now let us consider the situation when we know u, δu, x and a certain interval $[y^-, y^+]$, and we wish to find such $(v, c) = (\bar{v}, \bar{c})$ that the interval $[y^-, y^+]$ would be the interval of maximal error with respect to y at the point

$(x,\bar{y}) = (x(u,\bar{v}), y(u,\bar{v}))$. For this purpose we have to solve the following system of equations with respect to $v = (v_1, v_2)$ and c:

$$y^- = y^-(u, \delta u, v, c)$$
$$y^+ = y^+(u, \delta u, v, c) \tag{6.5}$$
$$x = x(u, v)$$

By (6.2) the system of equations (6.5) can be reduced to the equivalent one with two unknowns y and c

$$y^- = y^-(u, \delta u, v(u; x, y), c)$$
$$y^+ = y^+(u, \delta u, v(u; x, y), c) \tag{6.6}$$

Before we introduce further assumptions let us denote:

$$x_i = x(u_i, v_i), \qquad y_i = y(u_i, v_i)$$
$$c_i = \delta v_{i1} = \delta v_{i2}, \qquad i = 1, 2, \ldots, n. \tag{6.7}$$

Assumptions:

f) for every $i = 1, 2, \ldots, n$ there is given an interval $[y_i^-, y_i^+]$ such that, the system of equations:

$$y_i^- = y^-(u_i, \delta u_i, v(u_i; x_i, y), c)$$
$$y_i^+ = y^+(u_i, \delta u_i, v(u_i; x_i, y), c) \tag{6.8}$$

has exactly one solution $(y, c) = (\bar{y}_i, \bar{c}_i)$ which satisfies conditions

$$y_i^- < \bar{y}_i < y_i^+, \quad \bar{c}_i > 0; \tag{6.9}$$

g) there exist continuous functions $u: R \to R^k$, $\delta u: R \to R^k$ such that

$$u_i = u(x_i), \quad \delta u_i = \delta u(x_i) \tag{6.10}$$

If for each $i \neq j$, $i, j = 1, 2, \ldots, n$; it occurs that $x_i \neq x_j$, then the *assumption* g) is always fulfilled since, for example, simple broken lines can stand for $u(x)$ and $\delta u(x)$. It has to be mentioned, however, that even if the *assumption* g) is not fulfilled, the method can still be successfully applied. This assumption has only been introduced for the sake of the easier description of the method and its geometrical interpretation.

Let $f(x;a)$ and $c(x;z)$ be continuous real functions, $x \in R$, $a \in R^p$, $z \in R^q$. If variables a and z are treated as parameters, we obtain two families of functions $f(x;a)$ and $c(x;z)$.

By replacing in the formulae (6.2), (6.3) and (6.4) $u = u(x)$, $\delta u = \delta u(x)$, $y = f(x;a)$, $c = c(x;z)$, $x \in I = [\min\{x_i\}, \max\{x_i\}]$, we get the following functions:

$$f^-(x;a,z) = y^-\big(u(x),\delta u(x), v\big(u(x); x, f(x;a)\big), c(x;z)\big)$$
$$f^+(x;a,z) = y^+\big(u(x),\delta u(x), v\big(u(x); x, f(x;a)\big), c(x;z)\big) \tag{6.11}$$

For fixed $(a,z) = (\bar{a},\bar{z})$ the set

$$Z(\bar{a},\bar{z}) = \Big\{(x,y): x \in I, f^-(x;\bar{a},\bar{z}) \le y \le f^+(x;\bar{a},\bar{z})\Big\} \tag{6.12}$$

will be called the zone of maximal error along the function $f(x;a)$, $x \in I$, (Fig. 6.3).

It is easy to see that for any $x_0 \in I$ the interval $Z(\bar{a},\bar{z}) \cap \{(x_0,y): y \in R\}$ is the interval of maximal error with respect to y at the point $(x_0, f(x_0;\bar{a}))$ (Fig. 6.3).

6.2.3   Description of the method

In what follows we assume that all the assumptions formulated in Sect. 6.2.2 are fulfilled. The presented method comprises three stages.

First stage. We start with marking on a plane all the experimental points (x_i,y_i). Their coordinates are to be calculated from (6.7). Next, to get the general idea about how the zone of maximal error looks like, we take a few comparatively simple functions $f(x)$, typical for the analyzed problem, and a few functions $c(x)$ representing the expected magnitude of experimental error, and employ them, pair by pair, to draw by (6.11), the respective zones of maximal error. Usually this provides us with enough data to decide on the best shape of the proposed zone of maximal error Z along the unknown fitting function $f(x)$ on the whole interval I, taking care of all the experimental points. As a result we come up with a hand-drawing (Fig. 6.1) of two functions $f^-(x)$ and $f^+(x)$, $x \in I$, being the lower and the upper limitations of the zone of maximal error Z. At last, we read from that drawing the values $y_i^- = f^-(x_i)$ and $y_i^+ = f^+(x_i)$ for each x_i and, if useful, for other arbitrary points from the interval I.

Second stage. Having got the values y_i^- and y_i^+ from the first stage, we solve the system of equations (6.8) for $i = 1, 2, \ldots, n$; consequently, we obtain (Fig. 6.2) two auxiliary sequences of points $(x_i,\bar{y}_i)$, and $(x_i,\bar{c}_i)$. For the

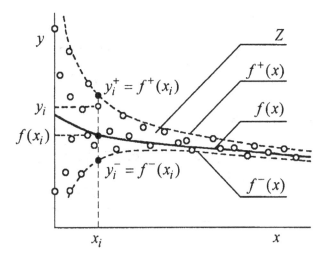

Figure 6.1. Asymmetry of the experimental errors along the *f(x)* function

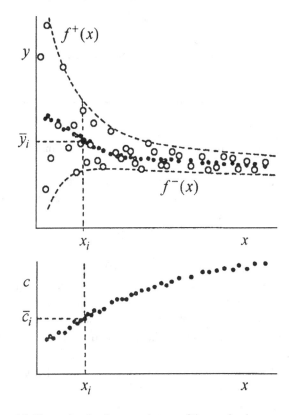

Figure 6.2. Illustration for the second stage of the maximal errors method

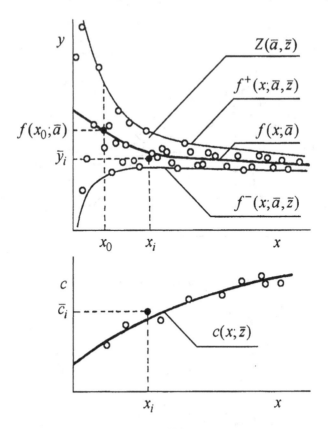

Figure 6.3. Zone of maximal error along the function $f(x;\bar{a})$ determined in the third stage

points $(x_i,\bar{y}_i)$ there is no more asymmetry and there occurs a considerably smaller scatter than for (x_i,y_i), so they reflect more accurately the relation between x and y. Therefore, the determination of the form of $f(x;a)$ becomes much easier. Similarly, on the basis of the points $(x_i,\bar{c}_i)$, we can determine the form of the function $c(x;z)$ describing relations between x and c. Having established the families of the functions $f(x;a)$ and $c(x;z)$, by the method of the least squares we can determine the values of parameters $(a,z)=(a_0,z_0)$ for which the functions $f(x;a_0)$ and $c(x;z_0)$, $x \in I$, constitute the best fits of the points $(x_i,\bar{y}_i)$ and $(x_i,\bar{c}_i)$. When using the standard statistical methods it is possible to examine whether the assumed families of the functions $f(x;a)$ and $c(x;z)$ are correct. In the case when the form of the functions $f(x;a)$ and $c(x;z)$ is known, being suggested by physical premises, the second stage allows the initial calculation of the parameters of these functions.

Third stage. On the basis of the results of the second stage we know the families of the functions $f(x;a)$ and $c(x;z)$ appropriate for the problem under consideration, and then we calculate such values of parameters $(a,z)=(\bar{a},\bar{z})$ for which the following conditions are fulfilled:

i) for every $i = 1, 2, \ldots, n$, $(x_i, y_i) \in Z(\bar{a}, \bar{z})$,

ii) $F(\bar{a}, \bar{z}) = \min\{F(a, z) : a \in R^p, z \in R^q\}$,

where

$$F(a, z) = \sum_{i}^{n} \left(y_i - f^-(x_i; a, z)\right)^2 + \sum_{i}^{n} \left(y_i - f^+(x_i; a, z)\right)^2 \qquad (6.13)$$

The values $(\bar{a}, \bar{z})$ have also their geometrical interpretation, namely, the respective zone of maximal error $Z(\bar{a}, \bar{z})$ around $f(x; a)$, $x \in I$, is the narrowest among those which contain all the experimental points (Fig. 6.3).

6.2.4 Maximal error method and the least squares estimation

Let us assume that analogically to the least square method the magnitude of all errors is known and that no asymmetry of errors is observed. Therefore we know the values of $\bar{c}$ parameters and for any $i = 1, 2, \ldots, n$ the equations:

$$\begin{aligned} f^-(x_i; a, z) &= f(x_i; a) - \Delta y_i \\ f^+(x_i; a, z) &= f(x_i; a) + \Delta y_i \end{aligned} \qquad (6.14)$$

are satisfied. Here Δy_i denotes the maximal error corresponding to y_i at the point (x_i, y_i).

From (6.13) and (6.14) we get

$$\begin{aligned} F(a, z) &= \sum_{i=1}^{n} \left(y_i - f(x_i; a) + \Delta y_i\right)^2 + \sum_{i=1}^{n} \left(y_i - f(x_i; a) - \Delta y_i\right)^2 = \\ &= 2\sum_{i=1}^{n} \left(y_i - f(x_i; a)\right)^2 + 2\sum_{i=1}^{n} \Delta y_i^2 \end{aligned} \qquad (6.15)$$

and assuming

$$A = 2\sum_{i=1}^{n} \Delta y_i^2 \qquad (6.16)$$

we obtain

$$F(a, z) = 2\sum_{i=1}^{n} \left(y_i - f(x_i; a)\right)^2 + A. \qquad (6.17)$$

Therefore in the case when we know the value of all errors and there is no asymmetry of maximal errors the equation (6.13) is reduced to (6.17) where A is constant. This means that parameters a for which the function

$$F(a) = 2\sum_{i=1}^{n} (y_i - f(x_i;a))^2,$$ (6.18)

attains the minimum value are the same as in the least squares method.

In this way we have shown that basing on above mentioned assumptions the maximal error method reduces to the least squares method.

6.3   Calculation of the fitting function $r(\varepsilon)$

6.3.1   Method

We present the results obtained by using the method above, described for the determination of the relation between the coefficient of anisotropy r in cold rolled copper and the strain ε. In order to adapt the notations used in Section 6.2.2 to the considered problem, we assume:

$$x = \varepsilon, \quad u_{i1} = t_0, \quad v_{i1} = t_i, \quad \delta u_{i1} = c_0,$$
$$y = r, \quad u_{i2} = w_0, \quad v_{i2} = w_i, \quad \delta u_{i2} = c_0, \quad k = 2,$$
$$f(x;a) = r(\varepsilon;a), \quad f^+(x;a,z) = r^+(x;a,z), \quad f^-(x;a,z) = r^-(x;a,z).$$

The quantities to be used obtained as a result of direct measurements on the cold rolled copper [6.5] are given in Table 6.1, the calculated $r(\varepsilon)$ function – in Fig. 6.12.

The respective functions according to notations in Sect. 6.2.2 take the form:

$$\varepsilon = x(t_0, w_0, t, w) = \log \frac{t_0 w_0}{tw}$$ (6.19)

$$r = y(t_0, w_0, t, w) = \left(\log \frac{w_0}{w} \right) : \left(\log \frac{t_0}{t} \right)$$ (6.20)

We can easily check that for $w_0 > w > 0$, $t_0 > t > 0$ the above functions fulfil *assumptions* d) and f), (Chapter 6.2.2) and consequently

$$r^+(t_0, w_0, c_0, t, w, c) = \left(\log \frac{w_0 + c_0}{w - c}\right) : \left(\log \frac{t_0 - c_0}{t + c}\right), \tag{6.21}$$

$$r^-(t_0, w_0, c_0, t, w, c) = \left(\log \frac{w_0 - c_0}{w + c}\right) : \left(\log \frac{t_0 + c_0}{t - c}\right), \tag{6.22}$$

$$v_1(t_0, w_0, \varepsilon, r) = w_0 \exp\left(-\frac{\varepsilon}{1 + r}\right), \tag{6.23}$$

$$v_2(t_0, w_0, \varepsilon, r) = t_0 \exp\left(-\frac{\varepsilon r}{1 + r}\right). \tag{6.24}$$

Table 6.1. Experimental values (in mm) for the calculation of the $r(\varepsilon)$ function for the cold rolled copper; $z_0 = 30\%$, $\alpha = 67.5°/RD$, $t_0 = 5.00$ mm, $w_0 = 5.00$ mm, $c_0 = 0.005$ mm, $n = 41$

i	t_i	w_i	i	t_i	w_i	i	t_i	w_i
1	4.976	4.984	15	4.762	4.818	29	4.095	4.311
2	4.950	4.970	16	4.712	4.800	30	3.925	4.148
3	4.936	4.960	17	4.702	4.773	31	3.901	4.140
4	4.908	4.943	18	4.671	4.770	32	3.875	4.081
5	4.907	4.939	19	4.668	4.750	33	3.657	3.937
6	4.900	4.936	20	4.664	4.744	34	3.376	3.759
7	4.890	4.922	21	4.631	4.736	35	3.356	3.718
8	4.881	4.916	22	4.638	4.724	36	3.337	3.680
9	4.874	4.904	23	4.583	4.695	37	3.109	3.577
10	4.849	4.895	24	4.561	4.666	38	2.985	3.422
11	4.841	4.878	25	4.450	4.591	39	2.887	3.332
12	4.805	4.861	26	4.360	4.497	40	2.665	3.233
13	4.803	4.848	27	4.351	4.483	41	2.626	3.147
14	4.767	4.817	28	4.351	4.478			

It follows from the above relations that the system of equations (6.8) takes the form:

$$r_i^+ = \left(\ln \frac{w_0 + c_0}{w_0 \exp\left(-\frac{\varepsilon_i r_i}{1 + r_i}\right) - c_i}\right) : \left(\ln \frac{t_0 - c_0}{t_0 \exp\left(-\frac{\varepsilon_i}{1 + r_i}\right) + c_i}\right), \tag{6.25}$$

$$r_i^- = \left(\ln \frac{w_0 - c_0}{w_0 \exp\left(-\dfrac{\varepsilon_i \, r_i}{1 + r_i}\right) + c_i} \right) : \left(\ln \frac{t_0 + c_0}{t_0 \exp\left(-\dfrac{\varepsilon_i}{1 + r_i}\right) - c_i} \right). \tag{6.26}$$

The auxiliary sentences $(\varepsilon_i, \bar{r}_i)$ and $(\varepsilon_i, \bar{c}_i)$ determined by this system, have suggested the assumption of the following families of functions

$$r(\varepsilon, a) = a_1 \varepsilon + a_2 + a_3 \sqrt{\varepsilon^2 + a_4 \varepsilon + a_5}, \tag{6.27}$$

$$c(\varepsilon, z) = c_0 + z_1 \varepsilon + z_2 \varepsilon^2. \tag{6.28}$$

The values of parameters are:

$\bar{a}_1 = 0.457, \quad \bar{a}_2 = 0.747, \quad \bar{a}_3 = -0.568, \quad \bar{a}_4 = -0.282,$
$\bar{a}_5 = 0.0415, \quad \bar{z}_1 = 0.0175, \quad \bar{z}_2 = 0.0017.$

6.3.2   Procedure

The procedure aiming at the determination of the $r(\varepsilon)$ function from experimental data is the following (according to 6.2.3):

First stage. Taking advantage of theoretical solutions (Sect. 6.2.1, 6.3.2): and experimental results in investigation on the $r(\varepsilon)$ relationship for different polycrystalline metals [6.5, 6.6] and copper [100] single crystal [6.7] we calculate for some typical cases described by assumed $r = r(\varepsilon)$ relations the $r^-(\varepsilon)$ and $r^+(\varepsilon)$ functions; they delimit the range of experimental scatter at the assumed c_0 value of the initial experimental error [mm]. Therefore, if we assumed the macrohomogeneous deformation ($c = $ const. $= c_0$) and the linear relationship: $r = 1 - 0.05\,\varepsilon$ (similar to that observed in the annealed nickel [5.1] and annealed copper [5.2]) we obtained two pairs of relationships $r^-(\varepsilon)$ and $r^+(\varepsilon)$ for $c = c_0 = 0.010$ and $c = c_0 = 0.005$, respectively (Fig. 6.4). For the inhomogeneous deformation ($c \neq$ const.) the similar calculation has been performed for the same as above mentioned $r(\varepsilon)$ relationship and two different $c(\varepsilon)$ functions reproducing the inhomogeneous behaviour of two sorts of copper: medium grain size and coarse grain size ($c = 0.01 + 0.03\,\varepsilon$ and $c = 0.01 + 0.32\,\varepsilon$), respectively (Fig. 6.5).

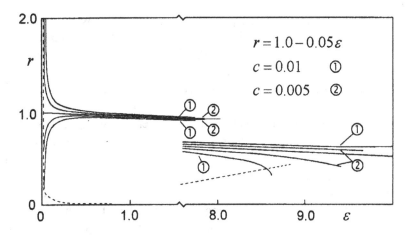

Figure 6.4. Effect of experimental error $c(\varepsilon)$ on the scatter of results along the function $r = 1.0 - 0.05\ \varepsilon$. Homogeneous deformation (c = const.) [6.2]

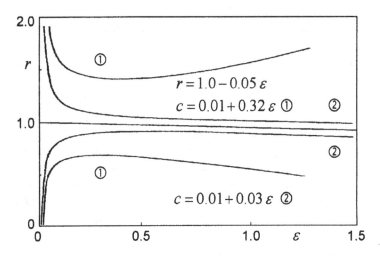

Figure 6.5. Effect of experimental error $c(\varepsilon)$ on the scatter of results along the function $r = 1.0 - 0.05\ \varepsilon$. Macroinhomogeneous deformation ($c = c(\varepsilon)$) [6.2]

In the $r(\varepsilon)$ diagrams (Figs 6.4 and 6.5) the area between $r^-(\varepsilon)$ and $r^+(\varepsilon)$ is the zone of experimental scatter for the given $r(\varepsilon)$ and $c(\varepsilon)$ functions. In the maximal errors method we take into consideration only these values of experimental w and t parameters for which

$$\frac{w_0 - c_0}{w_i + w_0} > 1 \quad \text{and} \quad \frac{t_0 - c_0}{t_i + w_0} > 1. \tag{6.29}$$

If the condition (6.29) would not occur, the results of measurement could not be used as leading to negative values of the strain ratio r. However, this phenomenon may have place only for very small ε values when Δw and Δt are smaller than the c_0 value. This observation leads to the appearance in the $r - \varepsilon$ diagrams (Figs 6.4 and 6.5) of the forbidden zone (marked with the dashed line).

When the elongation in the tensile test does not exceed the value of 2 (which is the case in real metals and alloys), the magnitude of scatter during the macrohomogeneous deformation tends to zero with rising elongation (Fig. 6.4). However, in macroinhomogeneous deformation the experimental scatter may attain its minimum much earlier (e.g. in the experiment presented in Fig 6.5 when $c = 0.01 + 0.32\,\varepsilon$), already at $\varepsilon = 0.35$.

The results obtained in this way give us the preliminary information on the shape of $r^-(\varepsilon)$ and $r^+(\varepsilon)$ functions, and facilitate the graphical determination of the limits of experimental scatter. Consequently, we mark on the plane all the experimental points (ε_i, r_i) and draw two curves $r^-(\varepsilon)$ and $r^+(\varepsilon)$, being the lower and the upper limitations of the zone Z of maximal error. Then, we read from that drawing the values $r_i^- = r^-(\varepsilon_i)$ and $r_i^+ = r^+(\varepsilon_i)$ for each ε_i and, if useful, for other arbitrary points within the whole interval I.

Second stage. Starting from the calculated in the first stage r_i^- and r_i^+ values we solve the system of equations (6.8) and (6.9) for $i = 1, \ldots, n$; consequently we obtain two sequences of points $(\varepsilon_i, \bar{r}_i)$ and $(\varepsilon_i, \bar{c}_i)$. For the points $(\varepsilon_i, \bar{r}_i)$ there is no more asymmetry and considerably smaller scatter than for (ε_i, r_i), so they reflect more accurately the relation between ε and r. On the basis of points $(\varepsilon_i, \bar{c}_i)$ we can determine the form of the function $c(\varepsilon; z)$ describing the relation between ε and c. As we have established that $r(\varepsilon)$ is a hyperbolic function and assumed for $c(\varepsilon)$ a parabolic function, by the method of least squares we determine the values of parameters $(a, z) = (a_0, z_0)$ for which the functions $r(\varepsilon; a_0)$ and $c(\varepsilon; z_0)$, $\varepsilon \in I$, constitute the best fits of the points $(\varepsilon_i, \bar{r}_i)$ and $(\varepsilon_i, \bar{c}_i)$.

Third stage. In the previous stage we established that: $r(\varepsilon; a)$ is a hyperbolic function; $c(\varepsilon; z)$ – a parabolic function and we know the values (a_0, z_0) – the first approximation of the sought for parameters (a, z). Availing ourselves of these data we have written a computer program for the calculation of values $(\bar{a}, \bar{z})$ such as:

i) for every $i = 1, 2, \ldots, n$, $(\varepsilon_i, r_i) \in Z(\bar{a}, \bar{z})$,
ii) $F(\bar{a}, \bar{z}) = \min\{F(a, z) : a \in R^5, z \in R^2\}$.

The function $F(a, z)$ and the zone of maximal error $Z(\bar{a}, \bar{z})$ along the $r(\varepsilon; \bar{a})$ function are described in detail in (Fig. 6.3).

6.3.3 Experimental verification on *f.c.c.* polycrystals

Some examples of the variation of the strain ratio of tensile tested poly-crystalline metals and the copper single crystal are presented below. They have been chosen from the preceding investigations, and are representative for different types of the $r(\varepsilon)$ characteristics and different magnitude of scatter.

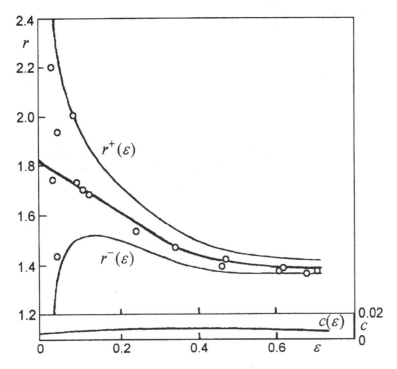

Figure 6.6. $r^{-}(\varepsilon)$, $r(\varepsilon)$ and $r^{+}(\varepsilon)$ relationships of the cold rolled 81/19 brass; $z_{rol} = 40.5\%$, $\alpha = 90°/RD$. Experimental data from [6.6]

Cold rolled brass: rolling reduction $r_{rol} = 40.5\%$, $\alpha = 90°/RD$ (Fig. 6.6). A rather sharp texture involves the considerably high $r_0 = 1.817$ value. The effect of straining is diminishing of r; at $\varepsilon \in [0.6, 0.7]$, r is stabilized at the value below $r = 1.4$.

Cold rolled brass: $z_{rol} = 40.5\%$, $\alpha = 0°/RD$ (Fig. 6.7) is characterized by a high anisotropy ($r_0 = 0.635$). Taking into account the form of the $r^{-}(\varepsilon)$ function calculated in the first stage (Figs 6.4 and 6.5) we omit in calcula-tions one experimental point. The straining induces the change of texture: at $\varepsilon = 0.7$ the strain ratio approaches the value 1, i.e. the condition of isotropy. In the $c(\varepsilon)$ function the increase of c with strain can be observed; $c_0 \approx 0.005$.

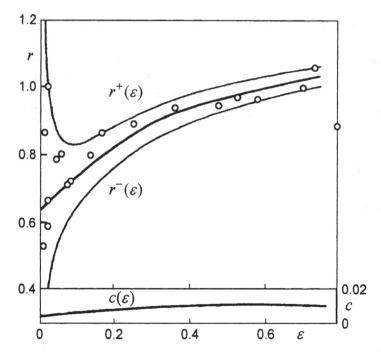

Figure 6.7. $r^-(\varepsilon)$, $r(\varepsilon)$ and $r^+(\varepsilon)$ relationships of the cold rolled 81/19 brass; $z_{rol} = 40.5\%$, $\alpha = 0°/RD$. Experimental data from [6.6]

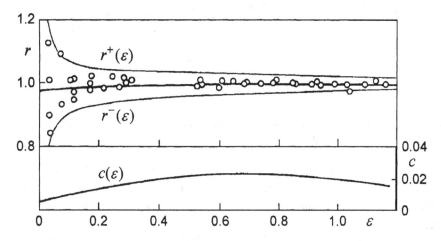

Figure 6.8. $r^-(\varepsilon)$, $r(\varepsilon)$ and $r^+(\varepsilon)$ relationships of the annealed 81/19 brass; $\alpha = 45°/RD$. Experimental data from [6.6]

Annealed brass: $\alpha = 45°/RD$ (Fig. 6.8). A close-to-random crystallographic orientation defines the state close-to-isotropy: $r_0 = 0.978$. The straining causes the subsequent diminishing of plastic anisotropy ($r = 0.995$

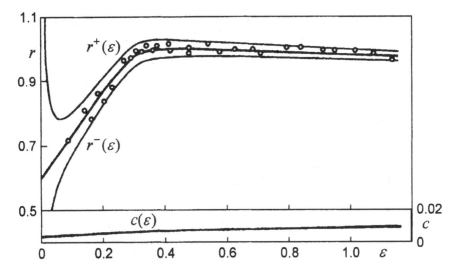

Figure 6.9. $r^-(\varepsilon)$, $r(\varepsilon)$ and $r^+(\varepsilon)$ relationships of the hot rolled nickel; $\alpha = 90°$/RD. Experimental data from [6.5]

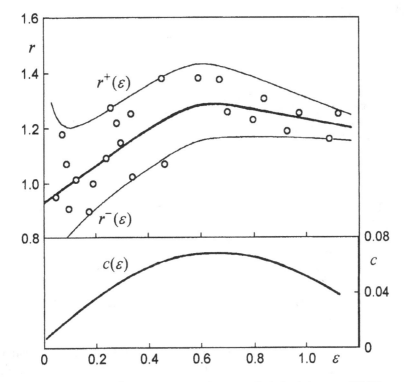

Figure 6.10. $r^-(\varepsilon)$, $r(\varepsilon)$ and $r^+(\varepsilon)$ relationships of the hot rolled aluminium; $\alpha = 90°$/RD. Experimental data from [6.5]

at $\varepsilon > 0.5$). The deformation is slightly macroinhomogeneous c varying from $c_0 = 0.005$ mm to $c = 0.022$ mm at $\varepsilon = 0.7$.

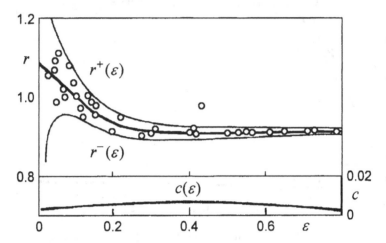

Figure 6.11. $r^-(\varepsilon)$, $r(\varepsilon)$ and $r^+(\varepsilon)$ relationships of the cold rolled 81/19 brass; $z_{rol} = 11\%$, $\alpha = 90°/RD$. Experimental data from [6.6]

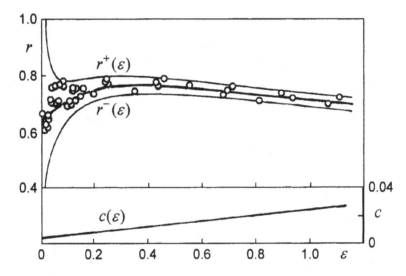

Figure 6.12. $r^-(\varepsilon)$, $r(\varepsilon)$ and $r^+(\varepsilon)$ relationships of the cold rolled copper; $z_{rol} = 30\%$, $\alpha = 67.5°/RD$. Experimental data from [6.5]

Hot rolled nickel: $\alpha = 90°/RD$ (Fig. 6.9). In the $r(\varepsilon)$ diagram two distinct zones are visible. $r_0 = 0.594$, $c_0 = 0.005$ mm; r rises with strain to attain the close-to-one value at $\varepsilon > 0.3$.

Hot rolled aluminium: $\alpha = 0°/RD$ (Fig. 6.10). The $r(\varepsilon)$ diagram is characterized by a large experimental scatter due to a coarse-grained structure. The

strong variation of c with strain evidences the macroinhomogeneous deformation. Initial value of the strain ratio $r_0 = 0.93$; in the $r(\varepsilon)$ diagram two
distinct zones with different slopes are visible.

Cold rolled brass: $z_{rol} = 11\%$, $\alpha = 90°/RD$ (Fig. 6.11). Taking into
account the general form of the $r^+(\varepsilon)$ function (Figs 6.4 and 6.5) one
experimental point distinctly departing from the zone of earlier described
distribution has been neglected. The deformation, is macroinhomogeneous;
$r_0 = 1.09$, $c_0 = 0.003$ mm.

Cold rolled copper: $z_{rol} = 30\%$, $\alpha = 67.5°/RD$ (Fig. 6.12). Two zones in
the $r(\varepsilon)$ function are visible with a continuous passing from first linear por-
tion ($\varepsilon \in [0, 0.1]$) to the second one ($\varepsilon > 0.35$), this description being similar
to this in Fig. 5.5. The variation of c with ε points to the macroinhomogeneity of deformation at the tension test; $r_0 = 0.63$, $c_0 = 0.005$ mm.

6.3.4   Verification on *f.c.c.* single crystals with stable crystallographic orientation

The *f.c.c.* single crystal with stable (in the tensile test) crystallographic
orientation (e.g. copper [100] single crystal characterized by $\alpha \leq \alpha_{cr}$ and
$\rho \leq \rho_{cr}$ – as described in Chapter 9) may serve for the experimental verification of the method of calculation of the $r(\varepsilon)$ function parameters, because it
behaves in a tensile test like a perfect crystal. The idea consists in the fact
that in the [100] *f.c.c.* single crystal $r(\varepsilon) = \text{const.} = 1.00$; the appropriate
parameters are those for which $r(\varepsilon)$ is the closest to the theoretically
determined course of the function $r(\varepsilon)$.

It has been mentioned in Sect. 6.2.1 that the zone of maximal error
should contain all experimental points being at the same time the narrowest
among those which possess this property. Therefore, for the proper descrip-
tion of the $r(\varepsilon)$ function such a value of c_0 should be chosen at which $r^-(\varepsilon)$
and $r^+(\varepsilon)$ functions would tightly embrace all experimental points. At the
same time only the first and the second ranges of the $r(\varepsilon)$ function described
by the hyperbola (5.1) should be taken into account (Chapter 3). But, as
detection of the $\varepsilon_{2/3}$ boundary may be experimentally difficult, the method
described below may be applied.

Let us calculate the $r(\varepsilon)$ function from the experimental data obtained at
the tensile testing of the 4-Cu [100] copper single crystal. Experimental results in the form (c_i, r_i) data (Table 7.2) and $r(\varepsilon)$ diagrams (Fig. 7.1) are
presented in the Chapter 7 which deals with the variation of the strain ratio
in deformed single crystals. We assume c_0 changing from $c_0 = 0.010$ mm,
$c_0 = 0.005$ mm to $c_0 = 0.004$ mm at constant strain range up to $\varepsilon = 0.12$. The
closest clasping of the experimental scatter by the $r^-(\varepsilon)$ and $r^+(\varepsilon)$ functions
and the smallest divergence of the r_0 value from the theoretical value

$r_0 = 1.000$ is observed at $c_0 = 0.004$ mm (Fig. 7.1); however, when the investigated range of straining is changing from $\varepsilon = 0.12$ to $\varepsilon = 0.90$ at constant $c_0 = 0.010$ mm, the best description by the hyperbola shape function is obtained when we take into account all experimental points up to $\varepsilon = 0.90$ (Fig. 7.2). Then the $r_0 = 1.012$.

The examined 4-Cu [100] sample fulfils the conditions of a "perfect single crystal" (as specified in Chapter 9). Therefore, the properties of a perfect single crystal should be: small value of c_0 and $r_0 = 1.000$. The statement of the accordance of the experimental value of r_0 with the theoretical one should be considered as the verification of the method of determination of anisotropy characteristics of single crystals.

6.4   Discussion

Using the proposed method of setting up the fitting function based on maximal errors, the procedure of determination of the $r(\varepsilon)$ function has been elaborated. The function is calculated from data characterized by asymmetric scatter, sometimes very large and varying with strain. The essential feature of the method is that it is based not on mean but on maximal errors. Its disadvantage is that an experimental point remarkably deviating from others can considerably affect the final result. To avoid this difficulty we introduced the first stage in the calculation of the $r(\varepsilon)$ function whose aim was the establishing of the form of $r^-(\varepsilon)$ and $r^+(\varepsilon)$ functions delimiting the zone of experimental scatter around the anticipated $r(\varepsilon)$ function. This permits to neglect one or two points distinctly departing from the zone of experimental scatter. Such a procedure has been successfully applied in three examples mentioned above (Figs 6.10 and 6.11).

In contrast to the method of the least squares, this method can be successfully applied even if we do not know the values of errors resulting from the measurement; moreover, these values can be determined in the process of approximation.

When comparing the values of r_0 calculated with the use of the presented method with those obtained by extrapolation of graphically interpolated $r(\varepsilon)$ functions, we come to the conclusion that the more anisotropic is the material the more useful is our way.

The proposed procedure is rather complex but gives univocal results and may serve for the calculation of new methods or new measuring devices.

6.5   References

6.1. W.Truszkowski et J.Kloch, *Modifications du coefficient d'anisotropie plastique en cours de déformation hétérogène*, Matériaux et Techniques, E 17 (1985).

6.2. J.Kloch, W.Truszkowski, *The Method of Determination of the Fitting Function Based on Maximal Errors*, Bull. Pol. Ac.: Techn., **34**, 683 (1986).

6.3. W.Truszkowski, J.Kloch, *Application of the Maximal Error Method for the Calculation of the r(ε) Function*, Bull. Pol. Ac.: Techn., **34**, 691 (1986).

6.4. W.Truszkowski, J.Piątkowski, J.Kloch, *Plastic Anisotropy in Strained Aluminium [110] Single Crystals*, To be published.

6.5. W.Truszkowski et J.Jarominek, *Essai de synthèse des recherches sur l'anisotropie plastique*, Mém. Sci. Rev. Métallurg., **70**, 433 (1973).

6.6. W.Truszkowski, J.Dutkiewicz et J.Szpunar, *Evolution de la texture et de l'anisotropie lors du laminage du laiton*, Mém. Sci. Rev. Métallurg., **67**, 355 (1970).

6.7. W.Truszkowski, J.Bonarski, *On the Imperfection of Crystallographic Orientation in FCC Single Crystals*, Z. Metallkd., **91**, (2000), in print.

Chapter 7

The variation of strain ratio in single crystals subdued to tensile test

7.1   Introduction

The relation of the strain ratio versus strain of $f.c.c.$ single crystals has been used in the elaboration of methods of the calculation of the $r(\varepsilon)$ function in mono- and polycrystalline materials. The criteria were the following: (i) the function properly describes the experimental data of the tensile test of high accuracy single crystals characterized by axial symmetry of slip systems and relatively small values of α and ρ parameters (α is the angle between the sample axis and the nominal [u v w] crystallographic direction, ρ is the half-width of the rocking curve). In perfect $f.c.c.$ single crystals with the [100] or [110] or [111] orientations the $r_{0\ theor}$ is equal to 1.000 or 0.500 or 1.000, respectively. The $r_{0\ theor}$ values have been calculated with the use of Krupkowski's method [7.1] considering the model of a perfect $f.c.c.$ lattice and taking into account the criterion of minimum work of deformation. We have shown previously [7.2, 7.3] that these conditions are fulfilled when $\alpha \leq \alpha_{crit}$ and $\rho \leq \rho_{crit}$; then the deformation of a single crystal at the tensile test is stable, i.e. $r(\varepsilon) = $ const. in the whole zone of uniform elongation. (ii) The parameters of the function $r(\varepsilon)$ have a well defined physical meaning.

The experimentally determined $r(\varepsilon)$ function of a single crystal which behaves at the tensile test like a perfect crystal can be used for the verification of the general method of calculation of the $r(\varepsilon)$ function parameters: such procedure is the best which leads to the least value of the parameter $|r_{0\ theor} - r_{0\ cal}|$ – Sect. 7.2.

7.2   Single crystals with stable crystallographic orientation

Copper single crystal was grown by the Bridgman technique and the tensile specimens with ca 6 mm diameter (precise values are given in Tables 7.1 and 7.2), and 20 mm gauge length were cut. Two [100] copper single crystal samples have been strained up to fracture [7.4]; their crystallographic characteristics was: 3-Cu [100]: $\alpha = 29'$, $\rho = 13'$; 4-Cu [100]: $\alpha = 7'$, $\rho = 16'$. The numerical results of tensile tests are presented in Tables 7.1 and 7.2 for both crystals, and in Figs 7.1 and 7.2 for the 4-Cu [100] sample, only.

Table 7.1. Tensile test results of 3-Cu [100] single crystal
Section AA: $w_0 = 6.147$ mm, $t_0 = 6.150$ mm
Section BB: $w_0 = 6.145$ mm, $t_0 = 6.136$ mm
Section CC: $w_0 = 6.137$ mm, $t_0 = 6.136$ mm

	AA		BB		CC	
	ε	r	ε	r	ε	r
1	0.0006	3.0049	0.0009	5.0122	0.0017	4.5040
2	0.0034	0.3126	0.0014	0.1251	0.0014	0.1249
3	0.0026	0.6006	0.0014	0.2862	0.0016	0.4285
4	0.0019	0.5005	0.0026	2.2057	0.0016	1.5007
5	0.0027	0.4170	0.0026	1.0021	0.0035	1.7514
6	0.0022	1.3355	0.0027	0.7013	0.0027	0.8891
7	0.0027	1.8368	0.0037	1.0933	0.0037	1.8767
8	0.0035	1.7534	0.0052	0.6008	0.0047	0.4497
9	0.0053	0.6505	0.0061	0.5840	0.0035	0.6923
10	0.0060	0.9486	0.0065	0.5390	0.0042	1.1672
11	0.0076	1.0451	0.0075	0.3941	0.0075	1.4224
12	0.0092	0.9011	0.0094	0.8138	0.0081	1.3823
13	0.0087	0.8007	0.0089	0.7197	0.0097	1.1436
14	0.0091	1.0015	0.0099	0.6493	0.0097	1.3090
15	0.0101	0.8797	0.0104	0.4546	0.0088	2.0036
16	0.0120	1.0573	0.0130	0.6672	0.0107	0.8332
17	0.0145	0.7804	0.0142	1.1782	0.0132	0.9286
18	0.0138	0.7349	0.0155	1.1622	0.0138	1.0242
19	0.0150	0.9587	0.0151	0.7555	0.0160	1.2807
20	0.0159	1.0871	0.0224	0.8275	0.0166	0.8210
21	0.0184	0.8530	0.0222	0.6790	0.0178	0.9122
22	0.0214	0.8993	0.0255	0.8806	0.0206	1.1009
23	0.0253	0.9385	0.0281	0.9562	0.0247	0.8871
24	0.0279	1.0876	0.0387	0.8588	0.0280	0.8996
25	0.0317	1.0659	0.0439	0.8288	0.0350	1.1533
26	0.0367	1.0382	0.0459	0.8987	0.0361	1.0377
27	0.0427	1.0015	0.0539	0.9595	0.0431	0.9850
28	0.0465	1.0529	0.0613	1.1121	0.0504	1.0272
29	0.0532	1.0597	0.0653	1.0177	0.0551	1.1575
30	0.0578	1.0073	0.0702	0.9460	0.0568	1.0364
31	0.0652	1.0382	0.0797	0.9936	0.0625	1.0221

Table 7.1. (continued)

	AA		BB		CC	
	ε	r	ε	r	ε	r
32	0.0679	1.0266	0.0853	0.9440	0.0709	0.9908
33	0.0758	0.9926	0.0912	0.9726	0.0792	1.1133
34	0.0841	0.9736	0.0941	0.9629	0.0817	1.0171
35	0.0907	1.0714	0.1003	1.0405	0.0889	1.0394
36	0.0890	0.9566	0.1092	0.9865	0.0973	1.0923
37	0.0981	1.0700	0.1159	1.0082	0.0974	1.0288
38	0.1051	1.0551	0.1201	1.0196	0.1027	1.0240
39	0.1135	1.0292	0.1338	1.0153	0.1162	1.0459
40	0.1213	1.0602	0.1441	1.0120	0.1335	0.9925
41	0.1325	1.0254	0.1525	1.0162	0.1420	0.9954
42	0.1402	1.0396	0.1639	1.0131	0.1533	0.9957
43	0.1526	1.0583	0.1783	1.0347	0.1619	1.0336
44	0.1633	1.0525	0.1983	1.0280	0.1773	1.0371
45	0.1771	1.0658	0.1980	1.0169	0.1993	0.9631
46	Neck		0.2007	1.0444	0.2031	0.9350
47	0.2181	1.1012				
48	0.2425	1.0754				
49	0.2753	1.0806				
50	0.3370	0.9970				
51	0.3956	0.9700				
52	0.4982	1.0228				
53	0.5881	1.0551				
54	0.7078	1.0467				
55	0.8796	1.0216				

Table 7.2. Tensile test results of 4-Cu [100] single crystal
Section AA: $w_0 = 6.137$ mm, $t_0 = 6.135$ mm
Section BB: $w_0 = 6.121$ mm, $t_0 = 6.122$ mm
Section CC: $w_0 = 6.135$ mm, $t_0 = 6.135$ mm

	AA		BB		CC	
	ε	r	ε	r	ε	r
1	0.0008	3.9964	0.0008	0.6663	0.0013	0.1427
2	0.0045	3.0000	0.0016	4.0007	0.0049	0.7643
3	0.0042	4.2007	0.0047	1.0712	0.0052	0.7774
4	0.0042	3.3333	0.0058	1.7701	0.0058	0.8944
5	0.0060	1.6422	0.0057	1.6930	0.0053	1.0624
6	0.0070	1.3881	0.0057	1.5004	0.0083	1.0399
7	0.0099	1.1777	0.0086	1.0382	0.0099	0.9058
8	0.0098	1.4998	0.0109	1.0937	0.0106	0.9694
9	0.0129	1.1343	0.0121	1.2428	0.0134	1.1583
10	0.0134	1.3426	0.0129	1.1947	0.0140	1.0977
11	0.0180	1.2447	0.0150	1.3009	0.0178	1.0184
12	0.0203	1.2141	0.0200	0.9358	0.0214	1.0470
13	0.0252	1.1688	0.0239	1.1163	0.0259	0.9746
14	0.0297	1.1291	0.0287	1.0836	0.0305	0.9998
15	0.0353	1.1501	0.0329	1.0408	0.0372	0.9647
16	0.0400	1.1507	0.0370	1.0838	0.0412	0.9998

Table 7.2. (continued)

	AA		BB		CC	
	ε	r	ε	r	ε	r
17	0.0502	1.0820	0.0422	1.0818	0.0493	1.0066
18	0.0544	1.1093	0.0506	1.0540	0.0545	0.9876
19	0.0601	1.0863	0.0554	1.0304	0.0597	1.0637
20	0.0655	1.0040	0.0627	1.0663	0.0672	1.0253
21	0.0712	1.0828	0.0687	1.0296	0.0726	0.9768
22	0.0731	1.0958	0.0728	0.9904	0.0746	1.0228
23	0.0788	1.0162	0.0764	1.0686	0.0795	1.0524
24	0.0855	1.0603	0.0820	1.0249	0.0865	1.0524
25	0.0916	1.0642	0.0866	1.0522	0.0907	1.0074
26	0.0981	1.0599	0.0939	1.0368	0.0984	0.9894
27	0.1043	1.0289	0.1009	1.0168	0.1034	1.0268
28	0.1113	1.0561	0.1056	1.0532	0.1107	1.0775
29	0.1160	1.0890	0.1118	1.0568	0.1162	1.0300
30	0.1301	0.9776	0.1185	1.0506	0.1313	0.9919
31	0.1411	0.9767	0.1314	0.9864	0.1386	0.9434
32	0.1474	0.9729	0.1423	0.9492	0.1451	0.9343
33	0.1574	0.9854	0.1484	0.9443	0.1535	0.9861
34	0.1638	1.0206	0.1576	0.9687	0.1634	1.0262
35	0.1734	0.9865	0.1687	1.0231	0.1700	0.9489
36	0.1773	0.9749	0.1753	0.9598	0.1775	0.9740
37	0.1825	0.9775	0.1809	0.9628	0.1814	0.9480
38	0.1862	0.9760	0.1868	0.9731	0.1835	0.9710
39	0.1875	0.9855	0.1903	0.9772	Neck	
40	0.1869	1.0458	0.1905	0.9754	0.2818	1.0645
41	0.1941	1.0175	0.1900	1.0187	0.2869	1.1224
42			0.1912	0.9755	0.3161	1.0803
43					0.3502	1.0628
44					0.3780	1.0284
45					0.4136	1.0155
46					0.4549	1.0016
47					0.4989	1.0340
48					0.5590	1.0288
49					0.6167	1.0530
50					0.6785	1.0553
51					0.7538	1.0390
52					0.8139	1.0180
53					0.8797	1.0255

It has been mentioned in Sect. 6.2.1 that the zone of maximal error should contain all experimental points being at the same time the narrowest among those which possess this property. Therefore, for the proper description of the $r(\varepsilon)$ function such value of c_0 should be chosen at which $r^-(\varepsilon)$ and $r^+(\varepsilon)$ functions would closely comprise all experimental points. At the same time only the first and the second ranges of the $r(\varepsilon)$ function described by the hyperbola (eq. 5.1) should be taken into account [7.5, 7.6]. But, as detection of the $\varepsilon_{2/3}$ boundary may be experimentally difficult, the method described in

Sect. 6.3.4 has been used. Then in the 4-Cu [100] sample the $r_0 = 1.012$, and the experimental error is 1.2 pct.

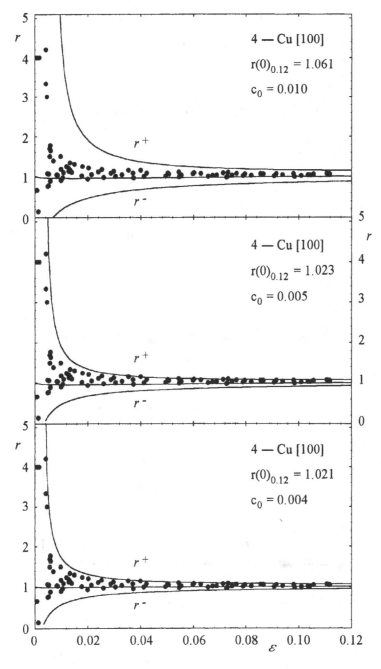

Figure 7.1. $r^-(\varepsilon)$, $r(\varepsilon)$ and $r^+(\varepsilon)$ relationships of the 4-Cu [100] copper single crystal. $c_0 = 0.010$; $c_0 = 0.005$; $c_0 = 0.004$

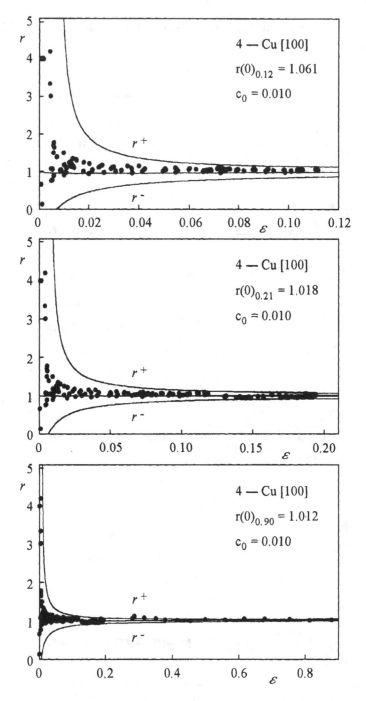

Figure 7.2. $r^-(\varepsilon)$, $r(\varepsilon)$ and $r^+(\varepsilon)$ relationships of the 4-Cu [100] copper single crystal. $c_0 = 0.010$

The conclusion: For the calculation of parameters of the $r(\varepsilon)$ function of the 4-Cu [100] single crystal experimental data of the whole zone of strain (up to $\varepsilon = 0.90$) should be considered, and the value $c_0 = 0.004$.

The examined 4-Cu [100] sample fulfils the conditions of a "perfect single crystal" (as specified above: $\alpha = 7'$, $\rho = 16'$): the $r(\varepsilon)$ function is constant within the whole zone of uniform elongation ($r = 1$). However, the obtained results have revealed that the small differences in the proposed c_0 values (0.010, or 0.005 or 0.004 – Fig. 7.1) and different ranges of the $r(\varepsilon)$ function taken for the analysis (from $\varepsilon = 0$ up to $\varepsilon = 0.12$, or $\varepsilon = 0.20$, or $\varepsilon = 0.90$ – Fig. 7.2) result in different r_0 values calculated. By varying the c_0 values we have shown that only when $c_0 = 0.004$ the zone of experimental scatter (between the $r^-(\varepsilon)$ and $r^+(\varepsilon)$ functions) contains all experimental points being at the same time the narrowest among those which possess this property. On the other hand, when changing the straining zone taken for the calculation we ascertain to explore only the first and the second range of straining (Sect. 3.3): passing in the calculations to the third range (above the value of strain $\varepsilon_{2/3}$ – Sect. 3.3) would result in a rapid change of the determined r_0 value, which was not observed. In this way we have shown that the zone $\varepsilon = 0$ to 0.90 contains the points only of the first and the second range. Concluding: in the determination of the proper value of r_0 in 4–Cu [100] sample, the $c_0 = 0.004$ should be assumed and all experimental results from the zone $\varepsilon = 0$ up to 0.90 must be considered.

The above described results should be considered as a physical basis of the proposed method of determining the parameters of the $r(\varepsilon)$ function.

7.3 Single crystals with unstable crystallographic orientation

The unstable behaviour of the strain ratio can be expected in the nominally symmetrical slip systems in strained cylindrical samples characterized by a considerable value of the ρ coefficient (texture in single crystal) or in other crystals with specially prepared non-symmetrical crystallographic orientation. The physical causes of instability of crystallographic orientation and of strain ratio, as well as the procedures aiming at its quantitative determination have been widely described in Chapter 9.

Below, some examples of the instability of nominally stable [100] and [110] and unstable [u v w] silver, copper and aluminium single crystals are presented, while the $r(\varepsilon)$ function was calculated using the procedure described in Sect. 6.3.2.

Single crystals of silver [7.7], copper [7.8] and aluminium [7.9] were grown by the Bridgman technique, and tensile specimens of 6 mm diameter and 50 mm length were cut; their nominal orientation was changing along

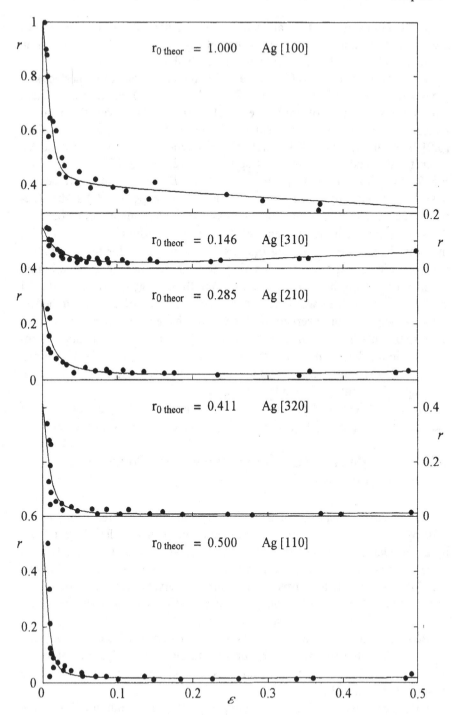

Figure 7.3. $r(\varepsilon)$ relations for tensile tested silver single crystals. Experimental data from the paper [7.7]; $r(\varepsilon)$ function recalculated using the equation (5.1)

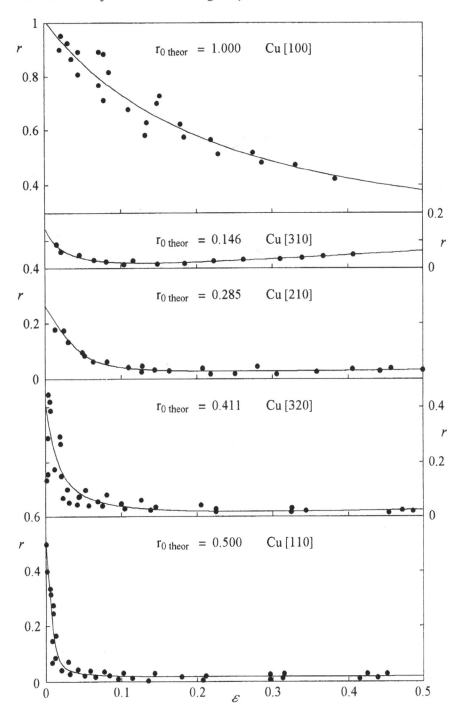

Figure 7.4. $r(\varepsilon)$ relations for tensile tested copper single crystals. Experimental data from the paper [7.8]; $r(\varepsilon)$ function recalculated using the equation (5.1)

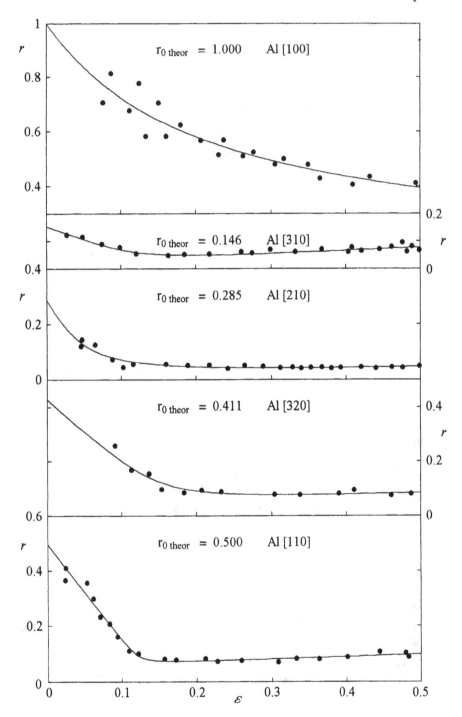

Figure 7.5. $r(\varepsilon)$ relations for tensile tested aluminium single crystals. Experimental data from the paper [7.9]; $r(\varepsilon)$ function recalculated using the equation (5.1)

the edge of the stereographic triangle: [100], [310], [210], [320] and [110]. It should be noted that all examined materials were *f.c.c.* metals with low (silver), medium (copper) and high (aluminium) stacking fault energy.

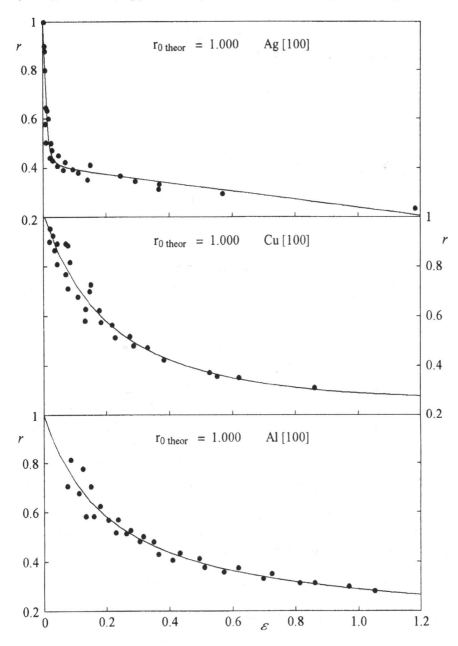

Figure 7.6. $r(\varepsilon)$ relations for tensile tested in a wide zone silver, copper and aluminium [100] single crystal [7.7, 7.8, 7.9]; $r(\varepsilon)$ function recalculated using the equation (5.1)

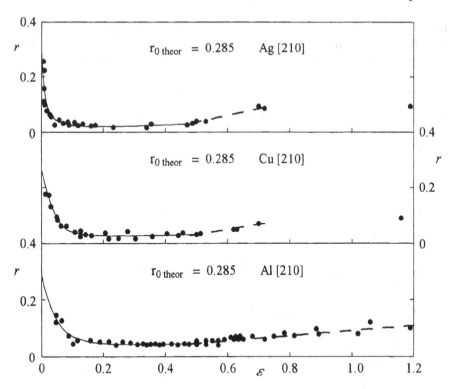

Figure 7.7. r(ε) relations for tensile tested in a wide zone silver, copper and aluminium [210]
single crystal [7.7, 7.8, 7.9]; *r(ε)* function recalculated using the equation (5.1) – full line, and
extended to the third range of strain – dashed line

The results are presented in Figs 7.3, 7.4 and 7.5 in the form of $r(\varepsilon)$
diagrams. As the main problem of this research was the experimental determination of the $r_{0\ cal}$ values and relating them to theoretically calculated
$r_{0\ theor}$, the zone of strain was limited to the first and second ranges (from $\varepsilon =$
0 up to $\varepsilon_{2/3}$)which, according to the previously proposed method can be
described by the hyperbola (eq. 5.1). Consequently, the parameters of the
$r(\varepsilon)$ equation for all fifteen samples were calculated for the zone from the
onset up to about 50 pct deformation.

It can be stated, however, that in [100] silver, copper and aluminium single crystals the second zone of strain extends up to the maximum elongation
(Fig. 7.6), while in [210] and [110] crystals (Figs.7.7 and 7.8) it ends at
about $\varepsilon = 0.5$ (i.e. $\varepsilon_{2/3} \cong 0.5$). In Figs 7.6 – 7.8 the $r(\varepsilon)$ function within the
first and the second ranges is described by equation 5.1 (full line); the further
deformation is continued in the third range (dashed line).

The macroinhomogeneity of all three examined metals can be inferred
already from the shape of the $r(\varepsilon)$ relation of [100] and [110] samples. In the
case of "perfect" [100] and [110] orientations (i.e. when $\alpha \leq \alpha_{crit}$ and the ρ

value is close to zero), the strain ratio would not change during the tensile test. However, the observed variation of the r value with strain at the onset of the deformation should be attributed to considerable values of both parameters: $\alpha > \alpha_{crit}$ and/or $\rho > \rho_{crit}$, and in consequence the deformation is macroinhomogeneous.

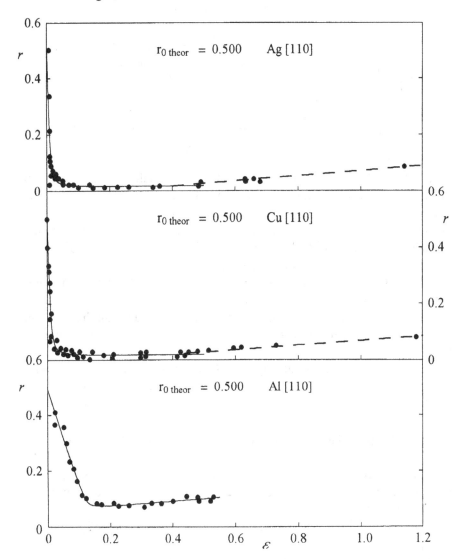

Figure 7.8. $r(\varepsilon)$ relations for tensile tested in a wide zone silver, copper and aluminium [110] single crystal [7.7, 7.8, 7.9]; $r(\varepsilon)$ function recalculated using the equation (5.1) – full line, and extended to the third range of strain – dashed line

The $r_{0\ theor}$ and $r_{0\ cal}$ ($r_{0\ cal}$ $= r(0)$) values for all examined samples are listed in Table 7.3. The results show that in spite of occasional large experimental scatter the difference between the theoretical and experimental r_0 values is negligible. In the author's opinion this is a strong argument for the physical meaning of the r_0 coefficient as an intrinsic material's property even when calculated from macroinhomogeneous deformation at the tensile test.

Table 7.3. Strain ratio r_0 values for *f.c.c.* single crystals

Crystallographic orientation	$r_{0\ theor}$	Silver	Copper	Aluminium	$\bar{r}_0$
[100]	1.000	1.014	1.012	0.999	1.008
[310]	0.146	0.152	0.146	0.153	0.150
[210]	0.285	0.291	0.264	0.288	0.281
[320]	0.411	0.419	0.391	0.428	0.413
[110]	0.500	0.518	0.508	0.494	0.507

From the point of view of the organization of crystallographic orientation in single crystals the author distinguishes three different states. (1) The model – perfect single crystal. It defines the limiting state which can be approached (but never attained) by improving the technique of the growing crystals. (2) Nominally stable (at the uniaxial tensile test), e.g. [100] crystal with $\alpha \leq \alpha_{crit}$ and $\rho \leq \rho_{crit}$, which in tensile test behaves like a perfect crystal. For the [100] copper single crystal the value α_{crit} was evaluated independently by Takeuchi [7.10] and Truszkowski and Wierzbiński [7.11] at the level of about 2°. The determination of the value ρ_{crit} is difficult; however, in the author's experiments it was possible to obtain stable behaviour in strained Cu [100] crystal sample when ρ was equal to about 20′. (3) Unstable (at the uniaxial tensile test) real single crystal (as well as nominally one stable with $\alpha > \alpha_{crit}$ and/or $\rho > \rho_{crit}$, as unsymmetrical, in relation to the crystal sample axis).

7.4   Calculation of the $r(\varepsilon)$ function in single crystals

The knowledge of the $r(\varepsilon)$ function of both, perfect and real, single crystals can be used for establishing a general method of the calculation of the $r(\varepsilon)$ function parameters.

There are several methods leading to precise determination of plastic anisotropy of materials; all are based on the strain ratio r versus strain ε relationship. In Chapter 3 a description has been proposed of the variation of plastic strain ratio during macroinhomogeneous deformation by a sequence of processes of homogeneous strain in which stable, unchanging deformation mechanism is operating. Two groups of procedures should be submitted for consideration [7.5]: the first, consisting in the determination of strain ratio at

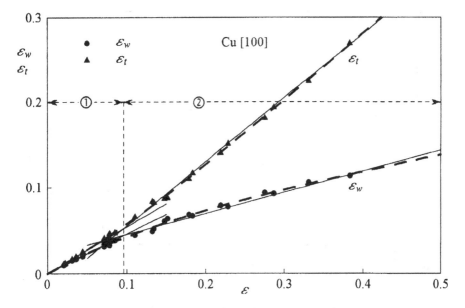

Figure 7.9. Two ranges in the $\varepsilon_w(\varepsilon)$ and $\varepsilon_t(\varepsilon)$ relationship of the Cu [100] single crystal described by two straightlinear functions, as in B-method (full line). Dashed line – hyperbolic description of the total zone, as in C-method. Experimental data from Truszkowski et al. [7.8]; Truszkowski and Kloch] [7.5]

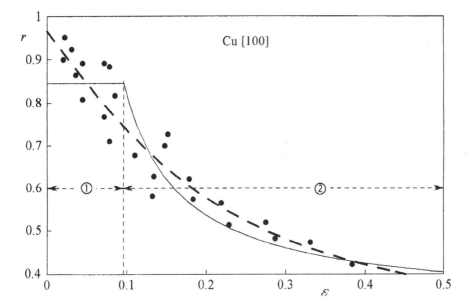

Figure 7.10. $r(\varepsilon)$ relations for the Cu [100] single crystal (as in Fig.7.9). Full line: $r(\varepsilon) = \varepsilon_w(\varepsilon)/\varepsilon_t(\varepsilon)$, where both partial strains are described by the straight lines (as in B-method). Dashed line: $r(\varepsilon) = \varepsilon_w(\varepsilon)/\varepsilon_t(\varepsilon)$, where both partial strains are described by the hyperbolic function (as in C-method)

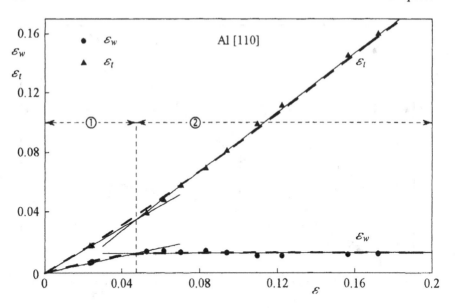

Figure 7.11. Two ranges in the $\varepsilon_w(\varepsilon)$ and $\varepsilon_t(\varepsilon)$ relationship of the Al [110] single crystal described by two straightlinear functions, as in B-method (full line). Dashed line – hyperbolic description of the total zone, as in C-method. Experimental data from Truszkowski et al. [7.9]; Truszkowski and Kloch] [7.5]

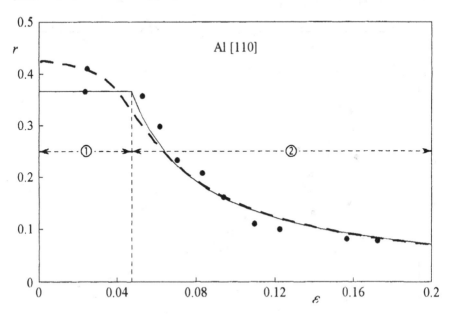

Figure 7.12. $r(\varepsilon)$ relations for the Al [110] single crystal (as in Fig.7.11). Full line: $r(\varepsilon) = \varepsilon_w(\varepsilon)/\varepsilon_t(\varepsilon)$, where both partial strains are described by the straight lines (as in B-method). Dashed line: $r(\varepsilon) = \varepsilon_w(\varepsilon)/\varepsilon_t(\varepsilon)$, where both partial strains are described by the hyperbolic function (as in C-method)

several levels of strain, and describing the fitting function in the whole first and second ranges by the hyperbola – method A, and the second one, when partial strains due to the change of width (ε_w) and thickness (ε_t) of a sample are related to the longitudinal strain ε – methods B and C.

Verification of the usefulness of the proposed methods was performed on [100] copper [7.8] and [110] aluminium [7.9] single crystals. The macroinhomogeneous deformation in the tensile test of both single crystals has found its expression in two segments of a straight line in the system ε_w and ε_t versus ε (Figs 7.9 and 7.11). In these crystals the unstable behaviour of the strain ratio (and consequently of the crystallographic orientation) was due to their imperfection. Two factors are responsible for this phenomenon in particular [7.12]. They can be included into the category of the "orientation errors": (i) deviation of the tensile test sample axis from the predetermined [u v w] orientation, and (ii) disorientation revealed in the structure of a single crystal yielding the effect of a strong, one-component texture, half-width of the rocking curve being its measure. The problem of the imperfections of real single crystals and their effect on unstable behaviour at the tensile test is widely discussed in Chapter 9.

Figures 7.9 and 7.10 illustrate the procedure of the determination of the $r(\varepsilon)$ relation in the Cu [100] single crystal in two steps [7.5]. In the first step, two straightlinear $\varepsilon_w(\varepsilon)$ fitting functions (and consequently two $\varepsilon_t(\varepsilon)$) – full line in Fig. 7.9 – allow to determine the boundary $\varepsilon_{1/2}$ between the ranges of plastic deformation ($\varepsilon_{1/2} = 0.096$). An optional solution for the first step is to describe both ranges by a hyperbolic equation (dashed line in Fig. 7.9). In the second step (Fig. 7.10) from the $\varepsilon_w(\varepsilon)$ and $\varepsilon_t(\varepsilon)$ relations we calculate the $r(\varepsilon)$ function in the range ①: $r(\varepsilon) = (a_1 \varepsilon)/(\beta_1 \varepsilon) = \text{const.}$, and in the range ②: $r(\varepsilon) = (a_2 \varepsilon + A_2)/(\beta_2 \varepsilon + B_2)$ – the hyperbolic function (full line in Fig. 7.10). We denote this procedure as method B (the thus calculated $r_0 = 0.845$). The optional description of the first step (i.e. the hyperbolic description of the $\varepsilon_w(\varepsilon)$ and $\varepsilon_t(\varepsilon)$ functions) gives for the whole zone of strain the $r(\varepsilon)$ relation in the form of the fourth order function (dashed line in Fig. 7.10). We denote this procedure as method C (the r_0 is here equal to 0.966).

The application of the above described mode to Al [110] single crystal gave for the A-method $r_0 = 0.494$, in the case of the application of the B-method: $\varepsilon_{1/2} = 0.047$ and $r_0 = 0.366$, and for the C-method $r_0 = 0.431$ (Figs 7.11 and 7.12) [7.5].

As in the B-method both $\varepsilon_w(\varepsilon)$ and $\varepsilon_t(\varepsilon)$ functions are linear in the first range ① and pass through the point $(0,0)$, their ratio has a constant value ($r = \text{const.}$); however, in the second range ② the straight lines $\varepsilon_w(\varepsilon)$ and $\varepsilon_t(\varepsilon)$ are not passing through $(0,0)$, their ratio is described by a hyperbola.

In the C-method it is not necessary to determine the $\varepsilon_{1/2}$ boundary between the first and the second ranges, and the total zone is described by the hyperbolae $\varepsilon_w(\varepsilon)$ and $\varepsilon_t(\varepsilon)$, their ratio giving the function of the fourth order.

Table 7.4. Strain ratio r_0 values for *f.c.c.* single crystals

Sample	$r_{0\ theor}$	Method A	Method B	Method C
Cu [100]	1.000	1.012	0.845	0.966
Al [110]	0.500	0.494	0.366	0.431

It follows from the above analysis and from the results obtained on Cu [100] and Al [110] single crystals listed in Table 7.4 that the most accurate results (i.e. deviating least from theoretically calculated values) are the r_0 data calculated with the use of the A-method.

7.5   General conclusions: single crystals and polycrystalline metals

There are several methods leading to precise determination of plastic anisotropy in materials. Broadly speaking, two groups of proceedings should be mentioned: the first one (method A) based on the determination of strain ratio at several levels of strain in the tensile test, and the second (B and C methods) when partial strains are related to the longitudinal strain.

It appears that the most versatile is the first one (method A). The r values are determined in a wide zone of tensile deformation from the smallest strain up to the limit of uniform elongation. The $r(\varepsilon)$ function is characterized by a considerable scatter of experimental data in a zone of small ε values, but the calculation of the relationship using the physically confirmed hyperbolic function (eq. 5.1) is based on experimental data of a zone of large strain in which the scatter of measured r values is much smaller. The extreme precision in the r value determination requires the use of a direct method; the indirect method ($r = -\varepsilon_w/(\varepsilon + \varepsilon_w)$) based on the assumption that the volume of the sample remains constant during deformation, defines the strain ratio as the relation of local width strain to the average thickness strain, and thus may constitute the source of additional error in the case of any inhomogeneity of the material. This method makes possible the precise determination of both, the r_0 value which qualifies the plastic anisotropy of material with a well defined physical meaning, and r_a – the strain ratio at a freely chosen strain (e.g. at the limit of uniform elongation) which is important for technology of plastic working (e.g. for deep drawing).

Using the A-method the present author determined in previous papers the r_0 values for several single crystals, e.g. for four [110] brass single crystals with different zinc content: 9.1; 10.0; 16.0 and 30.0 wt % Zn; the experimentally determined $r_{0\ [110]}$ values were $r_0 = 0.48$; 0.50; 0.55 and 0.50, respectively. The average value $\bar{r}_{0\,[110]} = 0.508$, which differs insignificantly from the theoretical value $r_{[110]} = 0.500$.

Finally, it is to be stressed that correct data are obtained even in the case of macroinhomogeneous deformation at the tensile test when two or more ranges appear in the $r(\varepsilon)$ relationship.

The second mode of describing the plastic anisotropy of materials (the B and C methods) could yield full and reliable characteristics of anisotropy only in the case of macrohomogeneous deformation of a strained sample, when within the whole zone of strain the same, unchanging deformation mechanism is operating. This, however, happens only in exceptional cases, e.g. in stable orientation single crystals or in mild steel. The analysis of the $\varepsilon_w(\varepsilon)$ function at the onset of straining of almost all materials tested by the author has shown at least two ranges of strain which is the evidence of the macroinhomogeneity of deformation. In these methods the first and the second ranges may be approached either by straightlinear $\varepsilon_w(\varepsilon)$ functions with different slopes (method B), or by a hyperbola passing through $(0,0)$ point (method C). The experimental evidence shows that the C-method gives results which are closer to reality.

7.6   References

7.1. A.Krupkowski, *Anizotropia mono- i polikrystalicznego metalu o strukturze A1*, Arch. Hutn., **2**, 9 (1957).

7.2. W.Truszkowski, S.Wierzbiński and A.Modrzejewski, *Influence of Mosaic Structure on Instability of the Strain Ratio in Deformed Copper Single Crystals*, Bull. Acad. Pol. Sci.,sér. techn., **30**, 367 (1982).

7.3. W.Truszkowski, *Quantitative Aspects of the Relation between Texture and Plastic Anisotropy*, Proc. Intern. Conf. on Textures of Marerials (ICOTOM 7), Noordwijkerhout, 723 (1984).

7.4. W.Truszkowski, J.Bonarski, *On the Imperfection of Crystallographic Orientation in FCC Single Crystals*, Z. Metallkd., **91**, (2000), in print.

7.5. W.Truszkowski and J.Kloch, *Effect of Inhomogeneous Tensile Deformation on Plastic Anisotropy*, Archives of Metallurgy, **45**, 3 (2000).

7.6. W.Truszkowski and J.Kloch, *New Aspects of Plastic Anisotropy in Materials*, Bull. Pol. Ac.: Techn., **46**, 289 (1998).

7.7. W.Truszkowski, J.Gryziecki and J.Jarominek, *Variation of Plastic Strain Ratio in the {001} Crystallographic Plane of Silver*, Bull. Pol. Ac.: Tech., **31**, 31 (1983).

7.8. W.Truszkowski, J.Gryziecki and J.Jarominek, *Variation of Strain Ratio in Cube Plane of Copper*, Metals Technology, **6**, 439 (1979).

7.9. W.Truszkowski, J.Gryziecki and J.Jarominek, *Assessment of the Strain Ratio in the Cube Plane of f.c.c. Metals*, Bull. Acad. Pol. Sci., sér. techn., **24**, 209 (1976).

7.10. T.Takeuchi, *Orientation Dependence of Work Hardening of Copper Single Crystals Near the [001] Axis*, J. Phys. Soc. Japan, **40**, 741 (1976).

7.11. W.Truszkowski, S.Wierzbiński, *Izmenene koefficienta plastičeskoj anizotropii monokristallov medi s orientacjej blizkoj [001] pri rastjazenii*, Fizika Metallov i Metallovedenie, **56**, 1195 (1983).

7.12. W.Truszkowski, *The Impact of Texture in Single crystals of FCC Metals on Mechanical Behaviour and Instability of Orientation*, Proc. Eight Intern. Conf. on Textures of Materials (ICOTOM 8), Ed. by Kallend and Gottstein, The Met. Soc., 537 (1988).

Chapter 8

Asymmetry of experimental scatter around the function $r(\varepsilon)$

8.1 Introduction

If, when evaluating plastic anisotropy of polycrystalline sheet metal, we assume that in formula (6.20) the width w_0 and w refer to the sample situated in the sheet plane with its axis at the angle θ to the rolling direction, and the thickness t_0 and t being measured in the direction perpendicular to the rolling sheet, then the coefficient r_θ describes the normal plastic anisotropy in a sample oriented at the angle θ to RD and strained to the elongation ε. The state $r = 1$ corresponds to the isotropy of the material, and the degree of anisotropy defines the deviation from this value, both in the range $r < 1$ and $r > 1$. On the other hand, the experimental scatter around the function $r(\varepsilon)$ is asymmetric; the change $0 < r < 1$ finds its equivalent in the change $1 < r < \infty$.

Approximation of the experimental results, based on the concept of maximal error of the value r, varying – as it occurs in the case of function $r(\varepsilon)$ – with the strain ε, allows to predict the scatter asymmetry around the analyzed function, at the assumed experimental error c_0. With known (or assumed) function $r(\varepsilon)$, the boundaries of scatter of the value of r are described by the functions $r^-(\varepsilon)$ (from below) and $r^+(\varepsilon)$ (from above).

The graphs in Figs 6.4 and 6.5 show that the experimental scatter around the function $r(\varepsilon)$, which is very great at small elongations, diminishes with macrohomogeneous deformation (when $c(\varepsilon) = \text{const.}$), and its asymmetry decreases simultaneously. On the other hand, during macroinhomogeneous deformation (when c varies with strain), the scatter of the value r, after exceeding certain strain, increases again, and is accompanied by the increase of asymmetry.

To reduce the asymmetry of scatter we can assume, as a measure of plastic anisotropy, the function $D(\varepsilon)$ [8.1]

$$D(\varepsilon) = \frac{r(\varepsilon)-1}{r(\varepsilon)+1} \tag{8.1}$$

and when representing anisotropy by a physically justified quantity r_0 (3.8) we obtain

$$D_0 = \frac{r_0 - 1}{r_0 + 1} \tag{8.2}$$

The functions $r(\varepsilon)$ and $D(\varepsilon)$ describe the change of plastic anisotropy of a material subjected to a tensile test. The values $r = 1$ or $D = 0$ correspond to the isotropy state; anisotropy is characterized by $r < 1$ and $r > 1$, or $-1 < D < 0$ and $0 < D < 1$.

8.2   Asymmetry of experimental scatter in copper [100] single crystals

Two single crystals 3-Cu [100] and 4-Cu [100] in the form of cylindrical samples of the dimensions: diameter of about 6.15 mm (accurate dimensions can be found in Tables 7.1 and 7.2) and the gauge length of about 20 mm, obtained by the Bridgman method, had the following characteristics: 3-Cu [100]: $\alpha = 29'$, $\rho = 13'$; 4-Cu [100]: $\alpha = 7'$, $\rho = 16'$. The samples were deformed on Instron testing machine at the strain rate 2 mm·min^{-1}, interrupting the tension to measure two diameters (w and t), perpendicular to each other, i.e. in the directions [010] and [001]. To do this, a toolroom microscope was used, with the measurement accuracy estimated at 0.002 mm. The results are listed in Tables 7.1 and 7.2. In the range of uniform elongation the change in deformation was measured in three sections AA, BB (in the middle of the sample) and CC, and after the instability limit only in the smallest section of the sample.

Geometrical analysis of a perfect single crystal of face centered cubic lattice of the orientation [100] shows that $r(\varepsilon) = \text{const.} = 1.0$. As it has been proved by the studies of Truszkowski et al. [8.2, 8.3] and Takeuchi [8.4, 8.5], the stability range of crystallographic orientation in the studied copper [100] single crystals is equal to $\alpha \le 2°$. Considering that for both samples, 3 and 4, the physical and geometrical conditions of stability, defined earlier [8.2, 8.3], are preserved, it can be assumed that the function $r(\varepsilon)$ approximating the experimental results of Cu [100] single crystals, subjected

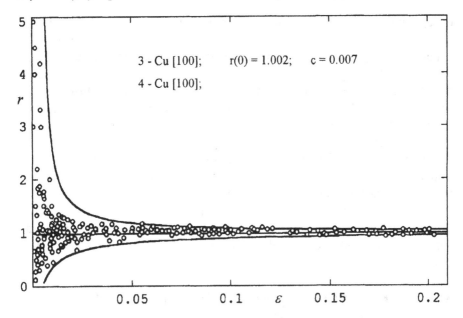

Figure 8.1. Functions $r^-(\varepsilon)$, $r(\varepsilon)$ and $r^+(\varepsilon)$ for copper single crystals 3-Cu and 4-Cu, with the orientation [100]

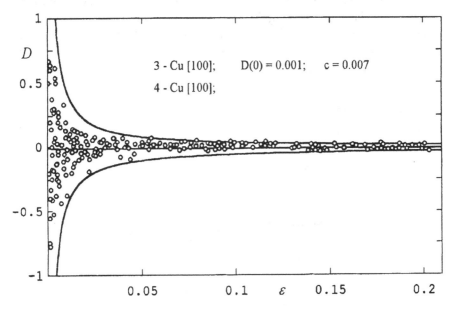

Figure 8.2. Functions $D^-(\varepsilon)$, $D(\varepsilon)$ and $D^+(\varepsilon)$ for copper single crystals 3-Cu and 4-Cu, with the orientation [100]

to tension, is $r(\varepsilon) = \text{const.} = 1.0$; thus $r_0 = 1.000$. Hence, any deviation from this value may be due only to incorrect estimation of the value of the experimental error c_0, assuming that experimental data for the calculations are taken only from the first and the second ranges (Chapter 3). From the experience, gained so far, it follows that under conditions existing at the time when tests of this type were carried out in the laboratory of the Institute for Metal Research of the Polish Academy of Sciences (now Institute of Metallurgy and Materials Science of PAS), the error in the measurement of c_0 in copper single crystals should be estimated at $c_0 = 0.004$ to $c_0 = 0.010$ mm.

The envelopes defining the range of scatter for the experimental results of both samples, 3 and 4 (Tables 7.1 and 7.2) and the approximating function $r(\varepsilon)$ were calculated using the method described in Chapter 6.

The results presented as graphs of the relations $r(\varepsilon)$ (Fig. 8.1) and $D(\varepsilon)$ (Fig. 8.2) confirm the usefulness of the method of maximal error and demonstrate the fully positive effect of symmetrization of the experimental scatter.

8.3   Quantitative approach to the asymmetry of experimental scatter

Let us consider the example of the function $r = 1.0 - 0.05\,\varepsilon$, discussed in Sect. 6.2. For the abcissa ε_i (Fig. 8.3) the range of experimental scatter of the ordinates is determined by the values r_i^- and r_i^+ as defined in (6.25) and (6.26), whereas the spacings of the point (ε_i, r_i) from the maximal and minimal deviations are as follows:

$$l_r^+(\varepsilon_i, r_i) = r_i^+(\varepsilon_i, r_i) - r_i,$$
$$l_r^-(\varepsilon_i, r_i) = r_i - r_i^-(\varepsilon_i, r_i). \tag{8.3}$$

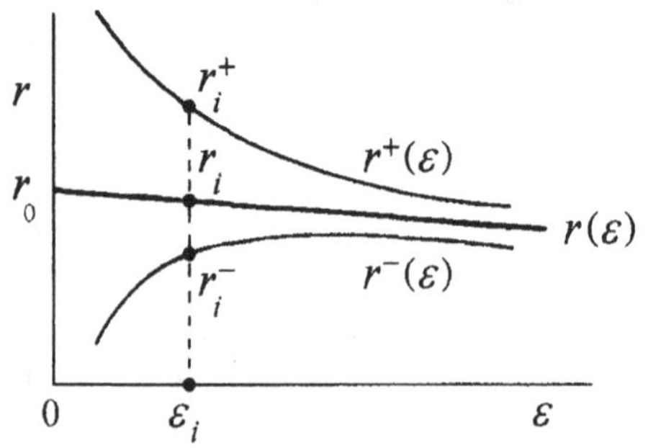

Figure 8.3. Determination of the coefficient A_r on the basis of the functions $r^-(\varepsilon), r(\varepsilon)$ and $r^+(\varepsilon)$

Let us assume $A_r(\varepsilon, r)$ based on the $r(\varepsilon)$ function as a measure of the asymmetry of experimental scatter

$$A_r(\varepsilon, r) = l_r^+(\varepsilon, r) - l_r^-(\varepsilon, r). \tag{8.4}$$

By analogy, the asymmetry of experimental scatter can be defined on the basis of the function $D(\varepsilon)$

$$A_D(\varepsilon, r) = l_D^+(\varepsilon, r) - l_D^-(\varepsilon, r) \tag{8.5}$$

where:

$$l_D^+(\varepsilon, r) = D(r^+(\varepsilon, r)) - D(r) \tag{8.6}$$

$$l_D^-(\varepsilon, r) = D(r) - D(r^-(\varepsilon, r)) \tag{8.7}$$

Having in view the confrontation of both measures of anisotropy ($r(\varepsilon)$ and $D(\varepsilon)$) it is reasonable to present the functions $A_r(\varepsilon, r)$ and $A_D(\varepsilon, r)$ in the same field.

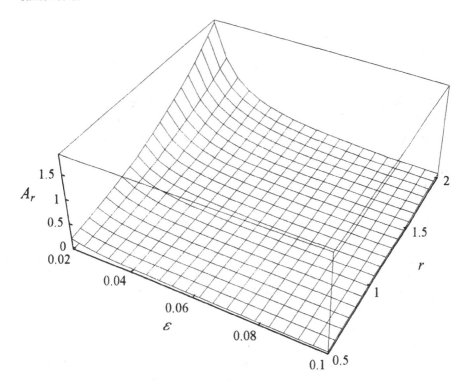

Figure 8.4. Function $A_r = f(\varepsilon, r)$ for macrohomogeneous deformation; $c_0 = $ const. $= 0.010$ mm

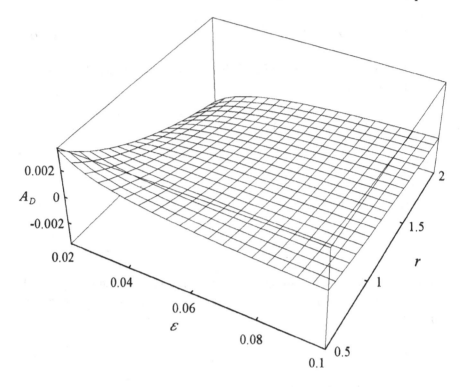

Figure 8.5. Function $A_D = f(\varepsilon, r)$ for macrohomogeneous deformation; $c_0 = $ const. $= 0.010$ mm

The concept of a quantitative description of the asymmetry of scatter around the $r(\varepsilon)$ function based on a model sample, was used with its properties similar to those analyzed experimentally in Chapter 6. Let us consider two cases: (a) macrohomogeneous deformation, and (b) macroinhomogeneous deformation.

a) Macrohomogeneous deformation: $w_0 = 6.15$ mm, $t_0 = 6.15$ mm, $c_0 = 0.010$ mm. The calculation results show the scatter asymmetry around the $r(\varepsilon)$ function A_r (Fig.8.4), and around the $D(\varepsilon)$ function A_D (Fig.8.5) for both functions in the same field (ε, r): $\varepsilon \in [0.01, 0.1]$ and $r \in [0.5, 2.0]$, which corresponds to: $\varepsilon = [0.01, 0.1]$ and $D \in [-1/3; 1/3]$.

Great asymmetry can be observed in the case of the description of the phenomenon by the function $r(\varepsilon)$, and a rather small asymmetry – by four orders smaller – in case of $D(\varepsilon)$.

During homogeneous deformation ($c = $ const.) the function $A_r = f(\varepsilon, r)$ (Fig. 8.4), at $\varepsilon = $ const. shows the increase of A_r with increasing r, which is distinctly visible at small value of ε (e.g. at $\varepsilon = 0.02$). On the other hand, a change of the degree of asymmetry A_D in the description of the phenomenon by the function $D(\varepsilon)$ (Fig. 8.5) shows an opposite tendency.

b) Macroinhomogeneous deformation: $w_0 = 6.15$ mm, $t_0 = 6.15$ mm, $c(\varepsilon) = 0.010 - 0.32\,\varepsilon$. From Fig 6.5 (in Chapter 6) it follows that the range of experimental scatter around the function $r = 1.0 - 0.05\,\varepsilon$, when $c = 0.01 - 0.32\,\varepsilon$, decreases with increasing deformation to about $\varepsilon = 0.3$, and then increases again. Hence, it appeared necessary to extend the investigated range of elongation to $\varepsilon = 0.5$.

Similarly, as in the case of macrohomogeneous deformation (Figs 8.4 and 8.5), also in the case of macroinhomogeneous deformation, passing from the measure $r(\varepsilon)$ to $D(\varepsilon)$ causes the diminishing of asymmetry of experimental scatter by about three orders of magnitude (Figs 8.6 and 8.7).

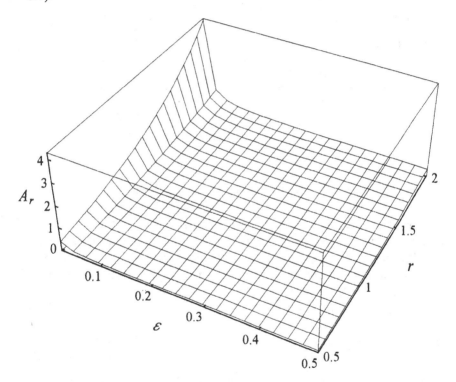

Figure 8.6. Function $A_r = f(\varepsilon, r)$ for macroinhomogeneous deformation; $c = 0.010 + 0.32\,\varepsilon$

The variation of the asymmetry of experimental scatter described by the functions $r(\varepsilon)$ and $D(\varepsilon)$, represented on model examples (Figs 8.4 – 8.7) suggests the activity of two mechanisms of deformation and the forming of texture, as the only factor, substantially affecting plastic anisotropy in single crystals and annealed single phase polycrystalline alloys. Description of the asymmetry phenomenon by means of the function $r(\varepsilon)$ demonstrates the operation of both macro- and micro-mechanisms, while the function $D(\varepsilon)$

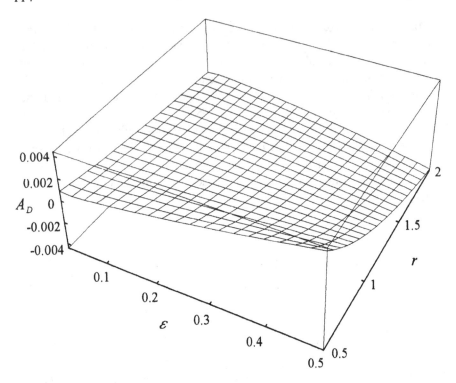

Figure 8.7. Function $A_D = f(\varepsilon, D)$ for macroinhomogeneous deformation; $c = 0.010 + 0.32\,\varepsilon$

eliminates the macromechanism. Since the macroeffect in the description of asymmetry by the function $r(\varepsilon)$ is $10^3 \div 10^4$ times stronger than the microeffect (transition from the function $r(\varepsilon)$ to the function $D(\varepsilon)$ diminishes the observed asymmetry 10^3 to 10^4 times), hence in the description of the phenomenon (Figs. 8.2 and 8.4) by the function $r(\varepsilon)$ this effect, presented graphically, is not observable.

8.4   Discussion

Asymmetric experimental scatter makes it difficult to calculate the function $r(\varepsilon)$ approximating the measurement results. For this reason it would be recommended to replace it by the function $D(\varepsilon)$ which, when describing the change in plastic anisotropy in the tensile test, reduces this disadvantageous feature to a great extent. This is accompanied by the positive phenomena: (1) the asymmetry of experimental scatter becomes radically reduced, and (2) the calculations are simplified: the function $D(\varepsilon)$ is approximated by the least square method.

At the same time, however, there appear effects which make the calculation difficult: (3) the straight line in the coordinates (ε, r) is translated into a hyperbola of the second order in the coordinates (ε, D). The rectilinear course of the relation $r(\varepsilon)$ has been confirmed in numerous experiments on single crystals of stable crystallographic orientation and polycrystals of random structure (or very weak texture, e.g. in mild steel); (4) the hyperbola of the second order in the system (ε, r) – which is a typical course for textured materials – changes into a "hyperbola of the fourth order" in the system (ε, D). A hyperbolic course of the relation $r(\varepsilon)$ in single crystals of metals with unstable crystallographic orientation has been observed experimentally. This effect occurred both in imperfect single crystals of the close to [100] and [110] orientations of copper, nickel and brass (i.e. those deviating from the nominal, stable orientation), as well as in those crystals in which the deviation from the state of a "perfect" single crystal was defined by the half-width of the orientation distribution.

However, it should be taken into consideration that all research institutions and industrial laboratories in the world describe plastic anisotropy on the basis of the coefficient r. The proposal of replacing this coefficient by the coefficient D could be recommended only in the case of unambiguous positive results.

8.5   References

8.1. J.A.Elias, R.H.Heyer and J.H.Smith, *Plastic Anisotropy of Cold Rolled-Annealed Low-Carbon Steel Related to crystallographic Orientation*, Trans. Met. Soc. AIME, **224**, 678 (1962).

8.2. W.Truszkowski, S.Wierzbiński, *Izmenene koefficienta plastičeskoj anizotropii monokristallov medi s orientacjej blizkoj [001] pri rastjazenii*, Fizika Metallov i Metallovedenie, **56**, 1195 (1983).

8.3. W.Truszkowski, *The Impact of Texture in Single Crystals of FCC Metals on Mechanical Behaviour and Instability of Orientation*, Proc. 8 th Intern. Conf. ICOTOM–8, Santa Fe, 537 (1988).

8.4. T.Takeuchi, *Orientation Dependence of Work Hardening of Copper Single Crystals Near the [001] Axis*, J. Phys. Soc. Japan, **40**, 741 (1976).

8.5. Y.Kawasaki and T.Takeuchi, *Cell Structures in Copper Single Crystals Deformed in the [001] and [111] Axes*, Scripta Met., **14**, 183 (1980).

Chapter 9

Instability of crystallographic orientation in real single crystals

9.1   Introduction

The perfect [u v w] single crystal in the form of a cylindrical sample is a body in which all atoms are in its lattice points and the [u v w] crystallographic direction is identical with the sample axis. It follows from this definition that such structure can never be attained in practice, as all real crystals are always burdened with the orientation errors. On the other hand the imperfection of metal single crystals has many aspects, all influencing their behaviour at uniaxial deformation, e.g. during the tensile test.

It is evident that in pure metals and single-phase undeformed polycrystalline alloys the crystallographic texture is the only parameter which has an essential effect on plastic anisotropy. Thus, it is possible to evaluate the variation of crystallographic orientation at the tensile test basing on the change of the strain ratio, on condition, however, that both, texture and anisotropy are referred to the same level of strain. Similarly, the experimentally determined and correctly described change of strain ratio in deformed single crystals can be applied to anticipate the variation of the crystallographic orientation.

In his earlier studies (Figs 7.3, 7.4, 7.5) the author has revealed a change in the strain ratio coefficient r with the strain ε in aluminium, copper and silver imperfect single crystals the sample axes of which were varying along the [100] – [110] edge of the standard stereographic triangle. It has been observed, but without stating precisely the actual deviation α of the crystallographic axis from the sample axis, that in the tested single crystals of the nominal [100] orientation the plastic elongation induces a sudden drop of the value of the coefficient r from $r_0 = 1$ at the initial state to a value below 0.3

in the range of considerable strain. On the other hand, in single crystals with the [110] nominal orientation this variation has been found to be considerably more intensive: from $r_0 = 0.5$ at $\varepsilon = 0$ to $r_a = 0.08$ in aluminium, $r_a = 0.05$ in copper, and $r_a = 0.03$ in silver strained up to the limit of uniform elongation.

These observations were at the origin of the method of the quantitative evaluation of instability of the strain ratio by means of the slope of the $r(\varepsilon)$ function measured at the onset of straining. It has been already mentioned in Chapter 7, that two factors are responsible for the phenomenon of instability at the uniaxial tensile test of nominally stable single crystals, in particular. They can be included in the category of the "orientation errors": (1) deviation of the single crystal axis from the predetermined "stable" orientation, expressed by the α angle, and (2) disorientation, revealed in the structure of the crystal, yielding the effect of a very strong, one-component texture, half-width of the rocking curve ρ being its measure.

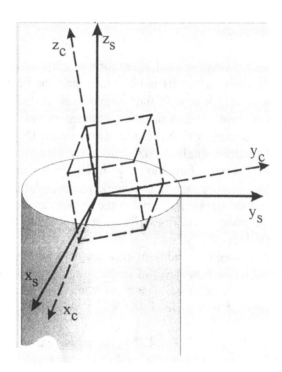

Figure 9.1. Disorientation between the co-ordinate systems connected with the sample (x_s, y_s, z_s) and with the crystal lattice (x_c, y_c, z_c) for a real single crystal sample

The α parameter is defined by the angle between the distinguished crystallographic orientation [u v w] and the axis of the sample; its finite value evidences the error of the machining of a tensile test specimen obtained from a single crystal. Test specimens are usually prepared from bigger single crystals grown by the Bridgman or the Czochralski method. As mentioned by Modrzejewski [9.1], it is especially desirable to use one matrix crystal in the case when a set of specimens differing slightly in crystallographic orientation is to be made. Growing of single specimens in succeeding runs might result in their different mosaic structure.

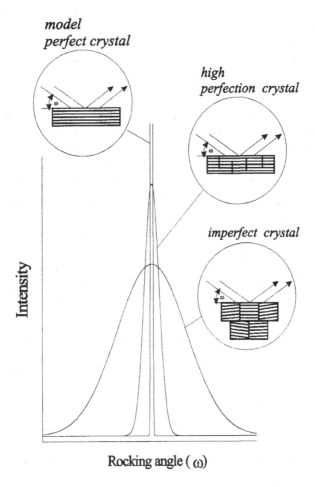

Figure 9.2. Simulated rocking curves for the model, high perfection and imperfect crystals, corresponding to different degrees of crystal perfection [9.2]

The nominal crystallographic orientation of the single crystal referred to the outward form (usually described by low Miller indices) often differs much from the actual orientation. This disorientation manifests itself in that the reference system connected with the sample does not agree with the system of the crystal (Fig. 9.1). In cylindrical samples the α parameter constitutes the essential criterion of the described disorientation.

The second parameter ρ characterises the imperfection of the inner structure of a crystal in the whole volume of the sample. The ρ value is described by the full width at half maximum (FWHM) of the rocking curve. The X-ray or neutron diffraction rocking curve is determined by measuring the diffracted intensity as the function of the rocking angle ω by which the crystal is rotated through its reflection position. Consequently, a real single crystal sample can be appraised as revealing a very strong, one-component texture; the diminishing of the sharpness of texture accompanies the appearance of the mosaic structure and results in the broadening of the rocking curve, as presented schematically in Fig. 9.2 [9.2].

Recently, several authors [9.1 – 9.4] were measuring the broadening of the rocking curve to follow the rising of the imperfection of crystal structure. A narrow rocking curve is characteristic for crystals of high perfection, while its broadening accompanies the formation of the mosaic structure [9.4]. From the point of view of its crystallographic orientation a single crystal could be considered as a perfect one when both parameters reach zero values: $\alpha = 0$, and $\rho = 0$. It is evident that the real single crystal is only a more or less accurate image of this model. In real single crystals the values of the α and ρ parameters depend on different factors, such as the method of their growing, the applied mechanical treatment, chemical and mechanical inhomogeneity, and may in extreme cases, number them among polycrystals.

The boundary between the criterion of a single crystal and that of a polycrystalline aggregate is conventional. Thus, a polycrystal with a very high ordering of crystallographic orientation, divided by low angle boundaries, may behave (in a certain respect, e.g. during plastic deformation), as a pseudo-single crystal. Baldwin Jr. [9.5] produced metal sheets of such properties by rolling copper to high deformation degrees prior to its annealing at high temperature. Roberts [9.6] has produced in copper a cube texture, a recrystallization texture in which cube planes lie parallel to the surface of the sheet with the cube direction parallel to the rolling direction. The texture is so sharp that the sheet may be regarded as a pseudo-single crystal. On the other hand, a real single crystal with developed mosaic structure may attain a high value of the parameter ρ classifying it to the group of polycrystals with very sharp texture. Since a change in the crystallographic orientation during uniaxial tension of single crystals and polycrystalline aggregates of annealed single phase alloys is strongly correlated with the change of the strain ratio, the instability of crystallographic orientation in the tensile test can be estimated by means of the change of the parameters of the $r(\varepsilon)$ function (ε is the true strain and r - strain ratio).

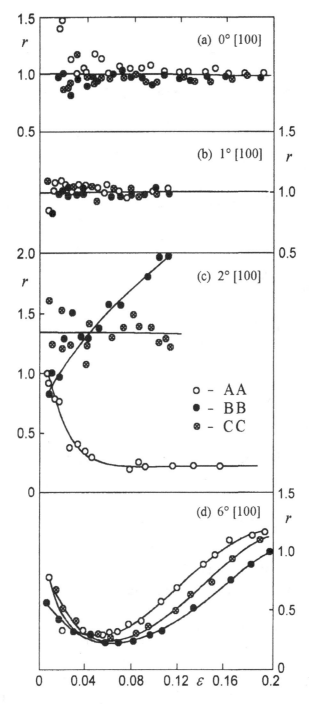

Figure 9.3. Variation of the strain ratio *r* at the tensile test of the *α* [100] copper single crystals in three sections: AA, BB and CC (experimental data: [9.7])

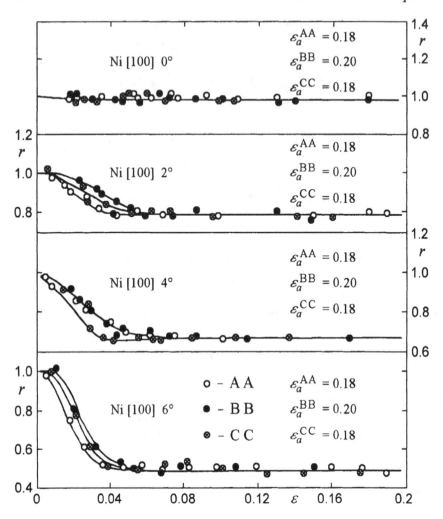

Figure 9.4. Variation of the strain ratio *r* at the tensile test of the *α* [100] nickel single crystals in three sections: AA, BB and CC (experimental data: [9.8])

9.2   Effect of the variation of the *α* parameter on the *r(ε)* relationship in single crystals

Experimental investigation was carried out on the near [100] copper single crystals [9.7]. The crystals were prepared from the 99.99% purity copper by the Bridgman method. Round section test pieces of the diameter $w_0 = t_0 \approx 6$ mm and 40 mm gauge length were used. The *α* angle was varying in the {110} plane from 0 up to 6 degrees (*α* = 0, 1, 2 and 6 degrees; the accuracy being $\Delta\alpha$ = 30 minutes). After mechanical working the samples were chemically polished in order to remove the deformed surface layer. The tensile test was carried out at room temperature with constant crosshead velocity 2 mm · min^{-1} and interrupted periodically for measurement of the contour of the section in three places at the length of the test piece: AA, BB (in the middle of the gauge length) and CC. The results exhibited perfect stability of the strain ratio at the tensile test up to α_{crit} equal to a value somewhat below 2 degrees (Fig. 9.3). It should be stressed that instability of the strain ratio is in this case equivalent to the instability of crystallographic orientation.

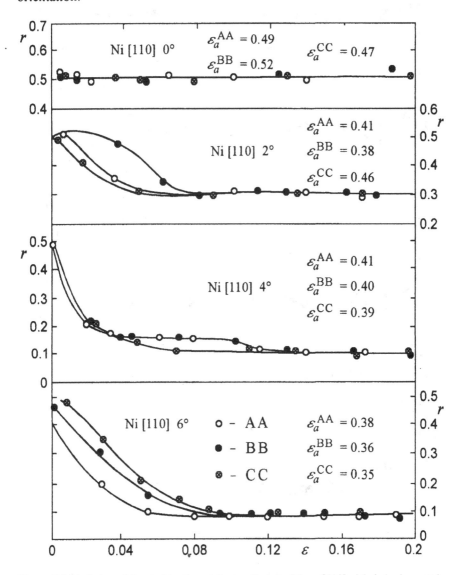

Figure 9.5. Variation of the strain ratio *r* at the tensile test of the α [110] nickel single crystals in three sections: AA, BB and CC (experimental data: [9.8])

However, nickel single crystals [9.8] with the nominal orientation [100], [110] and [111] prepared with the use of Czochralski method exhibited full stability when nominal value of α was equal to zero.

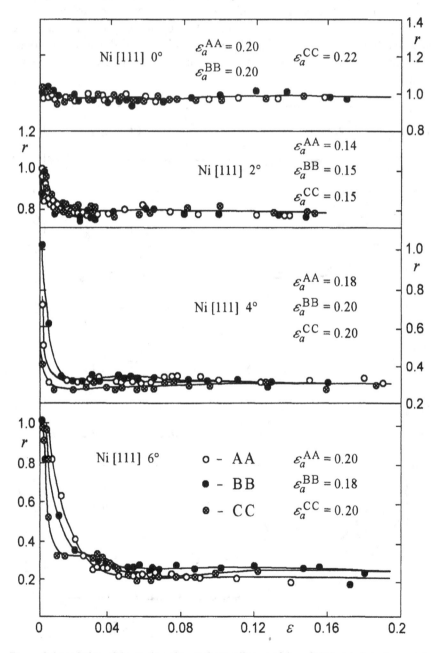

Figure 9.6. Variation of the strain ratio r at the tensile test of the α [111] nickel single crystals in three sections: AA, BB and CC (experimental data: [9.8])

Single crystals were prepared from 99.93% purity nickel by the Czochralski technique; rotation of seed was used to optimize homogenizing of the melt and to control the geometric configuration of the crystal. The method is capable of yielding high crystalline perfection. Round section samples of about 6 mm diameter and 20 mm gauge length were used. After mechanical working they were chemically etched and electrolitically polished. The tensile axes were oriented as follows: four single crystal samples with the deviation from [100]: $\alpha = 0°$, $2°$, $4°$ and $6°$ in the {001} plane, four samples with $0°$, $2°$, $4°$ and $6°$ deviation from [110] in the {001} plane, and four samples with $0°$, $2°$, $4°$ and $6°$ deviation from the [111] orientation in the {011} plane, in the direction towards [110]. The accuracy of the orientation of a sample axis was estimated at $0.5°$. The tensile test was carried out, as in the case of copper crystals.

The $r(\varepsilon)$ relationships for all examined nickel single crystals are presented in Figs 9.4, 9.5 and 9.6. Significant differences are observed between diagrams for AA, BB and CC sections, and this confirms that the deformation is macroinhomogeneous, at least in the zone of small strains. This inhomogeneity cannot be observed at $\alpha = 0$ in all the examined orientations ([100], [110] and [111]) as the $r(\varepsilon) = $ const. The experiments show that already a small rise in the α value causes the instability of the strain ratio at the tensile test.

In both series of experiments, in copper as well as in nickel single crystals, passing above α_{crit} caused the splitting of the samples: different parts of samples exhibited different mechanisms of deformation at the onset of straining. This phenomenon was observed already in 1976 by Takeuchi [9.9] in his research on the evolution of the work-hardening of copper single crystals with the tensile axes on the [001] – [102] line, at interval of 2 degrees. He has shown that "only the crystals which had the. [001] axis showed parabolic work-hardening curves, in which 8 slip systems equivalent with respect to the tensile axis worked uniformly. The specimens whose axes deviated by 2 ~ 6 degrees from the [001] axis split into two regions at the start of deformation: the first is the [001] – type region, and the second the [102] – type region in which one or two slip systems predominate."

Concluding: the described procedure enables the determination of the zone of crystals (from $\alpha = 0$ up to α_{crit}) in which the $r(\varepsilon)$ function is stable: $r(\varepsilon) = $ const. and consequently the crystallographic orientation of the single crystal does not change, on condition, however, that the value ρ is small ($\rho < \rho_{crit}$).

It should be stressed that in *f.c.c.* real single crystal samples with the nominal stable orientation ([100], [110] or [111]) deformed by straining, the range of stability of crystallographic orientation is rather small, hence precise determination of the α and ρ parameters is of considerable importance.

9.3   Effect of the change of ρ parameter on the $r(\varepsilon)$ relationship

As mentioned above (Sect. 9.1) the second factor having a significant effect on the instability of the strain ratio in deformed single crystals is the scatter of crystallographic orientation in the crystal (forming the mosaic structure), which in the author's investigations [9.10] was described by the half-width ρ of the neutron rocking curve.

Examination of a copper plate cut out from a large single crystal with a strongly inhomogeneous mosaic structure, made by Schneider [9.11] at different points by means of gamma rays diffraction technique, has shown that the local rocking curves representing the orientation of the mosaic blocks of the order of 1 μm reveal the half-widths of the orientation distributions varying from 4.7' to 1.1'. However, the average distribution function calculated from the reflexes for the whole single crystal was relatively smooth with half-width 10.5' – because of the spatial averaging over the crystal. In different parts of the crystal there occur numerous blocks of mosaic of weak orientation, or a fairly small number of blocks of stronger orientations; the particular areas are, moreover, misoriented with respect to each other, which is the criterion of the non-homogeneity of the mosaic structure, and consequently, of the "texture" of a single crystal.

In an earlier study [9.10] the authors traced the effect of the degree of crystallographic misorientation ("texture") in [110] copper single crystals. The crystals were prepared from 99.99% pure copper by the Bridgman method. Round section test pieces of the diameter $w_0 = t_0 \approx 6$ mm and 40 mm gauge length were used. After mechanical working they were chemically polished. All samples had the nominal [110] orientation with accuracy estimated at 0.5°. The tensile test was carried out at room temperature with constant cross-head velocity 2 mm·min^{-1}.

For experiments five single crystals have been chosen differing considerably in micromosaic structure described by the half-width of the neutron rocking curve: A: $\rho = 21'$, B: $\rho = 23'$, C: $\rho = 24'$, D: $\rho = 40'$ and E: $\rho = 47'$.

The $r(\varepsilon)$ relationships for each examined single crystals are presented in Fig. 9.7. Significant differences are observed between diagrams for AA, BB and CC sections, and this confirms that the deformation is macroinhomogeneous; however, for every crystal the first part of the average $r(\varepsilon)$ diagram could be approached by a single relationship. The main difference in the behaviour of the examined samples is observed within the range of small strains, up to about $\varepsilon = 0.05$: the greater the ρ value of the mosaic structure, the more rapid is the decrease of r on the $r(\varepsilon)$ diagram at the onset of straining.

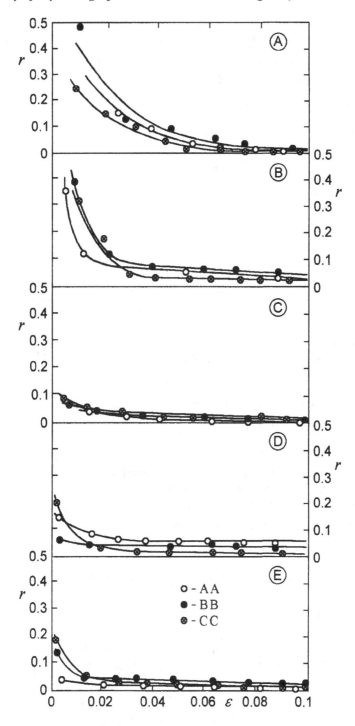

Figure 9.7. $r(\varepsilon)$ diagrams for five [110] copper single crystals in three sections AA, BB and CC (experimental data from [9.10])

In order to show clearly the difference in the instability of the strain ratio *r* (and consequently of the crystallographic [110] orientation) in the five examined single crystals, the first portions of the *r*(ε) diagrams of the A, B, C, D and E samples were put together in Fig. 9.8.

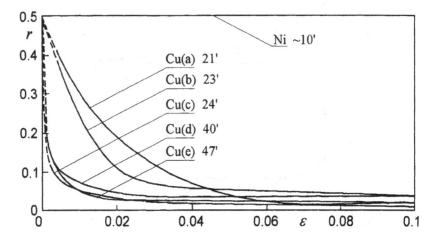

Figure 9.8. Influence of the *ρ* value in five [110] copper single crystals and in one [110] nickel single cristal on the *r*(ε) relationship (experimental data from [9.10] and [9.8], respectively)

The average *r*(ε) function for nickel [110] single crystal (with the crystallographic characteristics: nominal α value equal to zero and *ρ* value estimated at about 10') is presented in Fig. 9.8.

It should be stressed that both the mentioned errors, i.e. the difference between the nominal [u v w] and the real orientation, as well as the errors in the structure have a global effect on the imperfection of single crystals which in turn has an essential effect - as it has been shown in previous papers [9.7, 9.8, 9.10, 9.3] - on the mechanism of plastic deformation at the onset of the tensile test.

9.4   Evaluation of the α and ρ parameters

9.4.1   Determination of α

The orientation of the examined single crystal sample was close to [100] which implies appropriate distribution of directions in the plane perpendicular to its axis, as shown in Fig. 9.9. In order to define the sample orientation on a cut off fragment of the sample, the 200 back-reflection pole figure, measured in the space of the pole (χ) and azimuthal (ζ) angles with a grid of 0.05°× 0.05°, was registered [9.2]. Next, for the angular co-ordinates obtained in this way, the rocking curve was measured, which enabled to calculate accurately the crystal orientation in the sample system and to determine the disorientation value α.

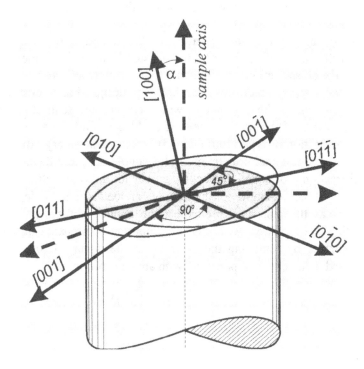

Figure 9.9. Disorientation between the crystal system (continuous line) and the sample system (broken line) with the nominal orientation [100]. The angle α is marked in the figure

Apart from the value of the α angle, the position of a disorientation plane, determined by the [100] crystal direction and the sample axis, and referred to the crystal system is also important. It influences the selection of the slip systems activated during plastic deformation of the anisotropic materials. Precise characteristics of the disorientation should contain these two elements, especially in the case of greater values of the α angle. In order to determine the position of the disorientation plane there have been carried out measurements of the rocking curve 200 and 020 on the side surface of the sample.

9.4.2   Determination of ρ

X-ray rocking curve is obtained by measuring the intensity of the reflection as a function of the rocking angle ω when the crystal (sample) is rotated through its Bragg position. The width of such a measured rocking curve

(FWHM parameter) is the measure of the angular orientation spread of the reflecting lattice plane. High perfection of the crystal is manifested by the narrow rocking curve, and a broader one is characteristic for lower perfection, caused by misoriented areas, described by the term "mosaic structure" - see Fig. 9.2.

The procedure of the determination of the α and ρ parameters was adjusted to the shape of the sample. It consisted in the measurement of reflection on the cross-section perpendicular to the sample axis and on the side surface of the sample. To do this there has been designed and constructed a special attachment to the goniometer which enabled the realisation of the planned algorithm of the measurement. It enabled to find the angular position of the reflection of type $\langle 100 \rangle$ and $\langle 110 \rangle$, connected by crystallographic relations with the nominal orientation of the sample axis, and then, to measure the diffraction profile and the rocking curves for these reflections.

Accuracy of the evaluation of the sample parameters α and ρ depends on such factors as the material's properties, the applied slit system and the wavelength of the radiation used. It is very important to prepare carefully the sample's surface by removing the outer layer, the damaged surface layer (DSL) caused by mechanical polishing with abrasive paper.

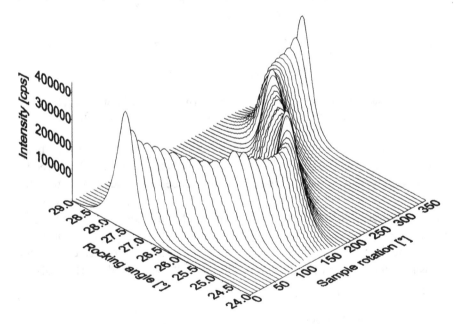

Figure 9.10. Set of rocking curves for 2-Cu [100] sample at different orientations during the rotation of the sample from 0 to 360°, at every 5°. CoKβ radiation was used [9.2]

According to Skorokhod [9.13] the DSL consists of the regions, which can be characterized by a greatly distorted surface layer with disoriented blocks and plastic deformation areas, located under the surface of the crystal. The elastic deformations of the surface regions relax very rapidly in the course of layer-by-layer etching. The thickness of the DSL may reach the dimension of six grain size of the abrasive paper used.

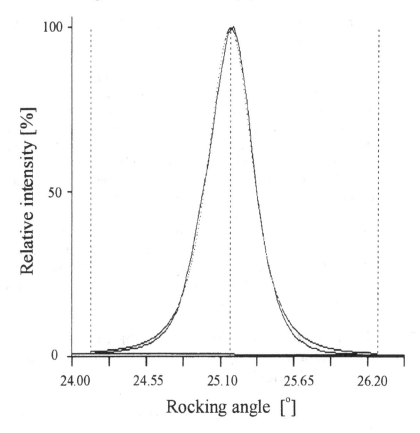

Figure 9.11. Single experimental rocking curve and its Lorentzian fitting function selected from a set of profiles presented in Fig. 9.10

9.4.3   Specification of the method on example of the [100] copper single crystal

The 2-Cu [100] single crystal cylindrical specimen of 10 mm diameter of the 99.99% purity copper, grown by the Bridgman procedure was used for the verification of the method [9.2]. Diffractometric measurements were made by means of 5-axis texture goniometer TZ6 equiped with a special single-crystal sample holder, using X-ray CoKβ ($\lambda = 1.62075$ Å), operated at 25 kV, 30 mA. The registered diffraction profiles were approximated by theoretical distributions, whose parameters were used to evaluate the degree of imperfection of single crystals. The criterion of the selection of the approximating function was the value of the smallest error of fitting the experimental data. The calculations were made using DAM [9.14] computer programme. The parameters of the selected distribution were accepted as the most reliable, on which the imperfections of the single crystals were assessed.

Figure 9.10 shows a set of rocking curves 200, registered for 2-Cu [100] crystal by means of CoKβ X-ray at various orientations of the sample. The successive orientations were obtained by rotating the sample with respect to the normal to its surface; in each position this normal being contained in the diffraction plane (determined by incident and diffracted beams).

The measurements have been carried out for orientation varying every 5° in the range of 0 to 360°. In this way two limiting orientations have been attained at the angular distance of 180°, for which the disorientation plane (containing α angle) was situated in the diffraction plane with the accuracy of 2.5° (Fig. 9.10). Rocking curves measured for these orientations of a sample provide information about α and ρ parameters. The first (α) as equal to the half of the variation range of the position of the curve; the second (ρ) as equal to the FWHM of the maximum intensity curve. Each registered rocking curve (Fig. 9.11) was separately analysed using the computer software package DAM [9.14]. This analysis makes possible the determination of the α angle and the average value of ρ. The numerical values for the examined 2-Cu [100] single crystal are: $\alpha = 0.65°$ (= 39'), $\rho = 0.267°$ (= 16').

9.5 The role of imperfection of *f.c.c.* single crystals in the analysis of instability of crystallographic orientation at uniaxial deformation

9.5.1 Calculation of α and ρ parameters in copper and brass single crystals

In the previous study [9.2] two [100] copper (3-Cu and 4-Cu) and one [100] brass single crystal samples have been investigated. The examined samples represented the medium (copper) and low (brass) stacking fault energy materials, previously [9.3] evaluated at $\gamma_{Cu} = 45$ mJ/m^2 and $\gamma_{Cu10Zn} = 26$ mJ/m^2.

Copper single crystals (made from 99.99 Cu purity) and brass (Cu10.0 wt % Zn) were grown by the Bridgman technique. For the tensile test specimens with ca. 6 mm diameter and 20 mm gauge length were prepared. The respective values giving the crystallographic characteristics of copper and brass single crystals are listed in Table 9.1 and Table 9.2.

The tensile test was carried out at room temperature with constant crosshead velocity 2 mm·min^{-1} and interrupted periodically for the measurement of the contour of the cross-section of the test piece at three (in case of the copper sample) or two (in case of brass) sections along its length.

Table 9.1. Crystallographic characteristics of Cu single crystals

Sample	Nominal orientation of the sample axis	Angular disorientation α			FWHM parameter p
		along the axis [100]	along the axis [010]	resultant value	200 reflection
3-Cu	[100]	$7 \pm 0.3'$	$28 \pm 0.3'$	$29 \pm 0.3'$	$13 \pm 0.3'$
4-Cu	[100]	$6 \pm 0.3'$	$2 \pm 0.3'$	$7 \pm 0.3'$	$16 \pm 0.3'$

Strain ratio describing plastic anisotropy is usually determined in the tensile test by measuring the variation of the dimensions of the cross-section of the strained test piece. As in the case of any experiment the results are burdened with error, its effect being the scatter of the calculated values. Theoretical considerations and experimental evidence [9.15 - 9.18] show that error in the r value and consequently the experimental scatter around the $r(\varepsilon)$ function is large in the zone of small strains and rapidly diminishes with the elongation. Additionally, remarkable asymmetry of scatter of experimental points with respect to the fitting function is observed.

Table 9.2. Crystallographic characteristics of Cu10.0 wt % Zn brass single crystal

Sample	Nominal orientation of the sample axis	Angular disorientation α	FWHM parameter p 200 reflection
Cu10Zn	100	evaluated $\alpha \leq 30'$	39'

When calculating the r coefficient from experimental data the sample is strained to considerable elongation, usually up to the onset of instability. The deformation is often producing a radical change of the texture and anisotropy; consequently the value of the coefficient obtained in this way cannot be considered as directly corresponding to the looked-for characteristics. The variation may be so great that the calculated strain ratio could no more be considered as the intrinsic property. In this situation the only solution of the problem of defining the plastic anisotropy with the well defined physical meaning appears the extrapolation of the $r(\varepsilon)$ function to the initial state, i.e. to $\varepsilon = 0$. Extensive studies of the author enabled to define the $(r - \varepsilon)$ dependence by a hyperbolic function with five constant parameters [9.18]. The advantage of this method, as it has been shown on the example of several *f.c.c.* single crystals and polycrystalline metals, is that the fitting function is based on the (ε_i, r_i) values measured at large elongation, where the experimental error in a correctly performed experiment is rather small. The coefficient $r_0 = r(0)$, obtained in this way, is an essential property of plastic anisotropy. The important feature of this method is that it is based not on mean, but on maximal experimental errors. Based on measurements in the particular steps of strain the $r^-(\varepsilon)$ and $r^+(\varepsilon)$ functions are determined, delimiting the zone of experimental scatter around the anticipated $r(\varepsilon)$ function. The described procedure gives univocal results, as it has been shown in many examples of single crystals and polycrystalline metals [9.16, 9.17].

9.5.2   Discussion

As the low values of the ρ parameters in both 3-Cu and 4-Cu single crystals are almost identical (Tables 7.1 and 7.2) and the α values do not exceed 30', the results have been put together in the one r vs ε diagram (Fig. 8.1). The fitted function $r(\varepsilon)$ can be approached by $r = \text{const.} = 1$. Characteristic is the large experimental scatter at low strain, and low scatter at the large deformation, above $\varepsilon = 0.05$. The analytically determined r_0 value, calculated on the copper crystal of the whole zone $\varepsilon = 0 \div 0.21$: $r_0 = r(0)_{0.21} = 1.002$. As the theoretical value is $r_0 = 1.000$, the experimental error in determining $r(0)$ is only 0.2 pct. (in spite of large scatter at the onset of straining).

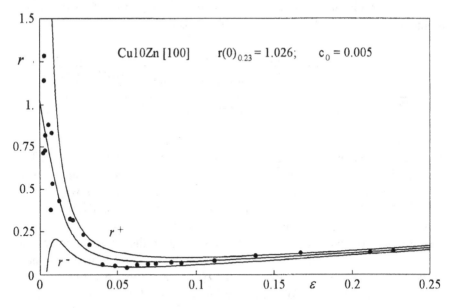

Figure 9.12. The $r(\varepsilon)$ function for Cu10Zn brass [100] single crystal with unstable crystallographic orientation [9.2]

Figure 9.12 represents the $r(\varepsilon)$ diagram for the whole zone of straining in the range of uniform elongation of the Cu10 wt % Zn single crystal with the nominal [100] crystallographic orientation. One can observe a large experimental scatter at the onset of straining, continuously diminishing with the deformation. Contrary to the case of copper single crystals, rapid drop in the

$r(\varepsilon)$ function of the examined brass is seen, testifying the instability of the r value during the straining. The $r(\varepsilon)$ function, calculated using the maximum error method [9.16, 9.17, 9.18] fits well the experimental results, and $r_0 = r(0)_{0.23}$ is equal to 1.026 (the experimental error in determining the r_0 value being 2.6%).

The results make possible the anticipation of the range of stability of crystals subdued to the uniaxial tensile test. The problem is of importance because the instability of crystallographic orientation in plastically deformed single crystals is at the basis of the process of formation of the deformation texture, both in single crystals and in polycrystalline metals

The results of the examination of the variation of the strain ratio at the uniaxial strain of all tested single crystals have shown that in spite of very large experimental scatter the calculated (on the basis of the earlier proposed method consisting in describing the $r(\varepsilon)$ function by the five parameters hyperbola [9.18]) function fits well the experimental data. While the scatter of the (ε_i, r_i) data in copper samples in the range of very small strain is as large as from $r = 0.12$ up to $r = 5.01$, the calculated value of $r_0 = \lim_{\varepsilon \to 0^+} r(\varepsilon) = 1.002$ is very close to the theoretical value $r_{0\,theor} = 1$.

A similar situation can be observed in the sample of the Cu10Zn brass single crystal deformed by straining. In spite of large experimental scatter, the calculated $r(\varepsilon)$ function fits well the experimental points and the $r(0) = 1.026$ value is close to the theoretical one: $r_{0\,theor} = 1$.

In medium stacking fault energy [100] copper single crystal, when α and ρ parameters have small values (evidently below α_{crit} and ρ_{crit}), one can observe the stable behaviour of the r value which is equivalent to the stable behaviour of the crystallographic orientation. However, in low stacking fault energy brass (with ten weight pct. zinc) with higher ρ value and strong tendency for unstable behaviour at the tensile test [9.3], the rapid drop at the $r(\varepsilon)$ relation, already at the onset of straining, manifests the instability of the strain ratio and of the crystallographic orientation.

9.6 A quantitative evaluation of instability of the strain ratio. The method

The variation of crystallographic orientation during plastic deformation of single crystals can be followed, as it has been suggested above, by the observation of the simultaneous change of the strain ratio. This suggestion justifies the concentration of the attention and the effort on the analysis of the $r(\varepsilon)$ function, i.e. on the variation of plastic anisotropy characteristics at the uniaxial tensile test of a single crystal as well as of single phase, recrystallized, polycrystalline metals and alloys.

In the case of the deformation of single crystals two extreme cases can be distinguished: (i) full stability of the coefficient r within the wide range of strain (at least in the whole zone of uniform elongation), characterized by the relation: $r(\varepsilon)$ = const., and (ii) instability of r during the plastic straining, and especially at the onset of the tensile test, where $r'(0)$ differs much from the zero value ($r'(0)$ describing the slope of the $r(\varepsilon)$ function at $\varepsilon = 0$).

The first case (i) can be illustrated by the example of the nickel single crystals, when the nominal value of α parameter is zero (Figs 9.4, 9.5 and 9.6), or of copper [100] $r(\varepsilon)$ function for 3-Cu and 4-Cu samples (Fig. 8.1). The second case (ii) is illustrated by examples of several single crystals of silver, copper and aluminium (Figs 7.3, 7.4 and 7.5) and the low stacking fault energy Cu10wt pct brass sample (Fig. 9.12).

It is proposed [9.3] to introduce the index φ as the measure of instability of the strain ratio at the uniaxial deformation, especially useful for the characterization of a single crystal:

$$\varphi = \frac{-r'(0)}{r_0} \tag{9.1}$$

where $r'(0)$ is the first derivative of the $r = f(\varepsilon)$ function at the point $(0, r_0)$ and $r_0 = r(0)$; for the orientations [100] and [110] in *f.c.c.* metals: $r_{0\,[100]} = 1.0$ and $r_{0\,[110]} = 0.5$ (Sect. 4.3.1).

As the author's early experimental results on fcc single crystals [9.19] suggested that there is a distinct influence of stacking fault energy on crystallographic orientation at the uniaxial deformation, a systematic investigation on brass single crystals have been undertaken.

9.7   Plastic anisotropy in brass single crystals

9.7.1   The effect of zinc on the anisotropy characteristics of alpha brass

To evaluate the effect of the stacking fault energy on the variation of the strain ratio and crystallographic orientation in tensile tested *f.c.c.* single crystals, four brass [110] crystals with zinc content varying from 9.1 up to 30 wt % were examined [9.19]. Single crystals were grown by the Bridgman method; they contained 9.1, 10.0, 16.0 and 30.0 wt % Zn, their half-width of the neutron rocking curve ρ was equal to 48', 39', 50' and 43', respectively, while the α value did not exceed 30'. Samples of the initial circular cross-section, and a diameter of about 6 mm, and the gauge length of 20 mm did not reveal any substantial change in their chemical composition along the length of a sample. Tensile test was carried out at room temperature at constant rate 2 mm · min⁻¹, when measuring the great axis *w* and the small axis *t* of the elliptical section of the strained sample in three sections: AA, BB (in the middle) and CC.

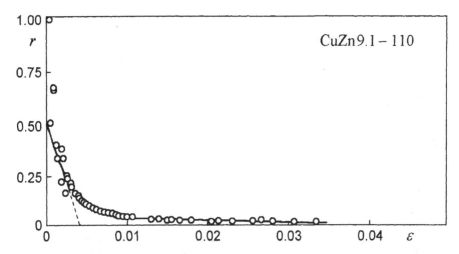

Figure 9.13. Variation of the strain ratio *r* at the tensile test of the [110] CuZn9.1 brass single crystal [9.19]

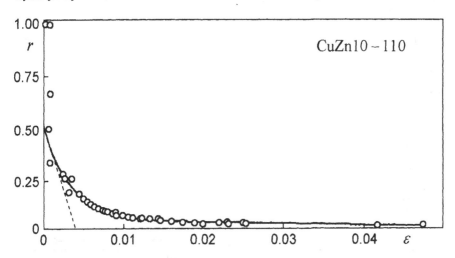

Figure 9.14. Variation of the strain ratio *r* at the tensile test of the [110] CuZn10 brass single crystal [9.19]

The results are presented in the form of the (ε_i, r_i) data for the whole zone of deformation at the tensile test (Tables 9.4 – 9.7), and the $r(\varepsilon)$ diagrams in the range of small deformations up to $\varepsilon = 0.05$ (Figs 9.13 – 9.16). When considering the scatter of the experimental points in the range of very small deformations (not exceeding $0.005\,\varepsilon$) the fitting function was calculated using the method of maximal error, developed by Truszkowski and Kloch [9.15 – 9.17]. In this way the values of r_0 have been calculated ($r_0 = r(0)$) for the examined brass crystals: CuZn9.1, $r_0 = 0.48$, CuZn10, $r_0 = 0.50$, CuZn16, $r_0 = 0.55$, CuZn30, $r_0 = 0.50$.

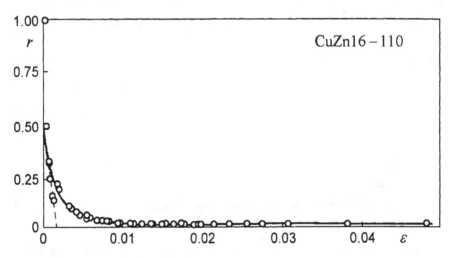

Figure 9.15. Variation of the strain ratio r at the tensile test of the [110] CuZn16 brass single crystal [9.19]

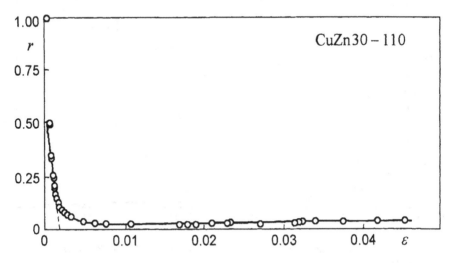

Figure 9.16. Variation of the strain ratio r at the tensile test of the [110] CuZn30 brass single crystal [9.19]

To calculate the values $r'(0)$ for CuZn9.1 and CuZn16 for which the $r(0)$ value was a little different from 0.5, the $r(\varepsilon)$ functions have been again determined with additional condition $r(0) = 0.5$. The results of calculations of the $r'(0)$ values are marked in Figs 9.13 – 9.16 by dashed lines and used for the calculation of the φ values (Table 9.3).

Table 9.3. Instability of strain ratio at the tensile test of the [110] *f.c.c.* single crystals

	Al	Cu	Ag	CuZn9.1	CuZn10	CuZn16	CuZn30
γ mJ/m^2	135	45	21.6	25.4	23.6	16.5	12.0
$\gamma/Gb \cdot 10^3$	18.2	4.33	3.05	2.44	2.27	1.59	1.15
$\varphi \cdot 10$	0.427	0.667	1.33	2.32	2.64	5.78	5.95
Reference	9.26	9.27	9.28	9.19	9.19	9.19	9.19

9.7.2 The effect of stacking fault energy on the instability of anisotropy characteristics in *f.c.c.* metals

In *f.c.c.* lattices as well as in *h.c.p.* ones the stacking fault energy is the essential factor in the investigation of plastic deformability of metals, the extent of which is related to the ratio γ/Gb (γ – stacking fault energy, G – shear modulus, and b – Burgers vector). It is widely known that stability of a close-packed lattice depends on the stacking fault energy, thus a strict dependence can be expected between the characteristics of instability of crystallographic orientation in deformed *f.c.c.* single crystals and the value of γ.

9.7.3 Determination of the stacking fault energy in alpha brasses

In the literature of the subject of the last forty years we find a wide spectrum of the values of the stacking fault energy. These divergences are particularly striking in the case of pure copper and its low-zinc alloys, and even more – in nickel and aluminium. Coulomb [9.20] made a list of γ_{Cu} values calculated by various authors using various methods, and revealed their extremely great scatter, from 30 to 163 mJ/m^2. On the other hand, a marked number of authors are ready to accept the value of the stacking fault energy of brass with high zinc content (CuZn30) to be close to 12 mJ/m^2. It seems that in the sixties and early seventies there was a general trend to overestimate the γ value, this tendency being stronger in higher stacking fault energy metals.

Taking advantage of the values of γ in alpha brasses with various content of zinc, measured in the years 1962 – 1971 by several authors [9.21 – 9.25], it was possible [9.3] to establish the $\gamma = f(x)$ dependence (x is zinc content), assuming $\gamma_{Cu} = 45$ mJ/m^2, as – according to Coulomb [9.20] the most probable value of the stacking fault energy of pure copper , and – suggested above – $\gamma_{CuZn30} = 12$ mJ/m^2. Hence for the calculation of the $\varphi = f(\gamma)$ relation the following values of γ have been proposed [9.3]: $\gamma_{Cu} = 45$ mJ/m^2, $\gamma_{CuZn9.1} = 25.4$ mJ/m^2, $\gamma_{CuZn10} = 23.6$ mJ/m^2, $\gamma_{CuZn16} = 16.5$ mJ/m^2, and $\gamma_{CuZn30} = 12$ mJ/m^2.

Table 9.4. Tensile test results of Cu9.1 wt % Zn [110] single crystal
Section AA: w_0 = 6.066 mm, t_0 = 6.065 mm
Section BB: w_0 = 6.054 mm, t_0 = 6.050 mm
Section CC: w_0 = 6.062 mm, t_0 = 6.064 mm

n	w	t	ε	r	sect.	n	w	t	ε	r	sect.
1	6.065	6.064	0.0003	1.000	A-A	34	6.063	6.024	0.0073	0.073	A-A
2	6.065	6.063	0.0005	0.500	A-A	35	6.063	6.022	0.0076	0.070	A-A
3	6.053	6.048	0.0005	0.500	B-B	36	6.059	6.021	0.0076	0.070	C-C
4	6.052	6.047	0.0008	0.666	B-B	37	6.051	6.005	0.0080	0.066	B-B
5	6.060	6.061	0.0008	0.667	C-C	38	6.059	6.018	0.0081	0.065	C-C
6	6.052	6.045	0.0012	0.400	B-B	39	6.059	6.015	0.0086	0.061	C-C
7	6.064	6.060	0.0012	0.400	A-A	40	6.063	6.014	0.0089	0.059	A-A
8	6.060	6.058	0.0013	0.333	C-C	41	6.063	6.011	0.0094	0.055	A-A
9	6.063	6.057	0.0018	0.375	A-A	42	6.063	6.008	0.0099	0.052	A-A
10	6.052	6.041	0.0018	0.222	B-B	43	6.051	5.990	0.0105	0.050	B-B
11	6.059	6.055	0.0020	0.333	C-C	44	6.051	5.975	0.0130	0.040	B-B
12	6.052	6.038	0.0023	0.166	B-B	45	6.063	5.984	0.0139	0.037	A-A
13	6.059	6.052	0.0025	0.250	C-C	46	6.059	5.977	0.0149	0.034	C-C
14	6.051	6.037	0.0026	0.230	B-B	47	6.051	5.960	0.0155	0.033	B-B
15	6.063	6.051	0.0028	0.214	A-A	48	6.059	5.968	0.0165	0.031	C-C
16	6.059	6.049	0.0030	0.200	C-C	49	6.051	5.945	0.0180	0.028	B-B
17	6.059	6.046	0.0035	0.167	C-C	50	6.051	5.930	0.0205	0.025	B-B
18	6.063	6.046	0.0036	0.158	A-A	51	6.063	5.940	0.0213	0.024	A-A
19	6.051	6.029	0.0040	0.143	B-B	52	6.051	5.915	0.0231	0.022	B-B
20	6.059	6.043	0.0040	0.143	C-C	53	6.051	5.900	0.0256	0.020	B-B
21	6.051	6.027	0.0043	0.130	B-B	54	6.061	5.910	0.0267	0.032	A-A
22	6.063	6.041	0.0045	0.125	A-A	55	6.059	5.900	0.0279	0.018	C-C
23	6.059	6.040	0.0045	0.125	C-C	56	6.051	5.885	0.0281	0.018	B-B
24	6.063	6.037	0.0051	0.107	A-A	57	6.051	5.878	0.0307	0.016	B-B
25	6.059	6.036	0.0051	0.107	C-C	58	6.050	5.855	0.0334	0.020	B-B
26	6.051	6.028	0.0055	0.100	B-B	59	6.059	5.620	0.0774	0.015	A-A
27	6.063	6.035	0.0055	0.100	A-A	60	6.059	5.593	0.0813	0.006	C-C
28	6.063	6.034	0.0056	0.097	A-A	61	6.055	5.333	0.1304	0.014	A-A
29	6.059	6.033	0.0056	0.097	C-C	62	6.049	5.285	0.1360	0.006	B-B
30	6.063	6.031	0.0061	0.088	A-A	63	6.059	5.286	0.1378	0.004	C-C
31	6.059	6.030	0.0061	0.088	C-C	64	5.970	3.769	0.4917	0.034	A-A
32	6.059	6.027	0.0066	0.081	C-C	65	5.977	3.558	0.5473	0.026	C-C
33	6.059	6.024	0.0071	0.075	C-C	66	5.976	3.930	0.7549	0.041	B-B

Table 9.5. Tensile test results of Cu10 wt % Zn [110] single crystal
Section AA: $w_0 = 6.116$ mm, $t_0 = 6.113$ mm
Section BB: $w_0 = 6.066$ mm, $t_0 = 6.064$ mm
Section CC: $w_0 = 6.065$ mm, $t_0 = 6.060$ mm

n	w	t	ε	r	sect.	n	w	t	ε	r	sect.
1	6.115	6.112	0.0003	1.000	A-A	37	6.062	5.994	0.0123	0.057	B-B
2	6.065	6.063	0.0003	1.000	B-B	38	6.062	5.987	0.0134	0.052	B-B
3	6.115	6.111	0.0005	0.500	A-A	39	6.112	6.031	0.0142	0.048	A-A
4	6.064	6.057	0.0007	0.333	C-C	40	6.062	5.981	0.0144	0.048	B-B
5	6.064	6.061	0.0008	0.666	B-B	41	6.061	5.976	0.0146	0.047	C-C
6	6.114	6.110	0.0008	0.666	A-A	42	6.061	5.967	0.0161	0.043	C-C
7	6.063	6.057	0.0008	0.666	C-C	43	6.061	5.959	0.0175	0.039	C-C
8	6.062	6.057	0.0010	0.999	C-C	44	6.111	6.003	0.0190	0.045	A-A
9	6.063	6.053	0.0023	0.272	B-B	45	6.061	5.950	0.0190	0.036	C-C
10	6.063	6.052	0.0025	0.250	B-B	46	6.061	5.944	0.0200	0.034	C-C
11	6.113	6.098	0.0029	0.200	A-A	47	6.062	5.936	0.0220	0.031	B-B
12	6.062	6.044	0.0031	0.187	C-C	48	6.111	5.979	0.0230	0.037	A-A
13	6.062	6.044	0.0031	0.187	C-C	49	6.061	5.925	0.0232	0.029	C-C
14	6.062	6.048	0.0033	0.250	B-B	50	6.111	5.967	0.0250	0.034	A-A
15	6.112	6.097	0.0033	0.250	A-A	51	6.062	5.916	0.0254	0.027	B-B
16	6.062	6.041	0.0045	0.174	B-B	52	6.061	5.817	0.0416	0.016	C-C
17	6.061	6.035	0.0048	0.160	C-C	53	6.062	5.788	0.0472	0.014	B-B
18	6.112	6.087	0.0049	0.153	A-A	54	6.111	5.816	0.0506	0.016	A-A
19	6.062	6.035	0.0055	0.138	B-B	55	6.061	5.709	0.0603	0.011	C-C
20	6.112	6.082	0.0057	0.129	A-A	56	6.062	5.663	0.0691	0.010	B-B
21	6.061	6.026	0.0063	0.117	C-C	57	6.111	5.664	0.0771	0.011	A-A
22	6.062	6.028	0.0066	0.111	B-B	58	6.061	5.499	0.0978	0.007	C-C
23	6.112	6.076	0.0067	0.108	A-A	59	6.062	5.432	0.1107	0.006	B-B
24	6.112	6.071	0.0075	0.095	A-A	60	6.111	5.427	0.1198	0.007	A-A
25	6.061	6.018	0.0076	0.095	C-C	61	6.061	5.288	0.1369	0.005	C-C
26	6.062	6.021	0.0078	0.093	B-B	62	6.062	5.200	0.1544	0.004	B-B
27	6.112	6.065	0.0085	0.083	A-A	63	6.111	5.189	0.1647	0.005	A-A
28	6.062	6.014	0.0089	0.080	B-B	64	6.040	4.916	0.2133	0.020	C-C
29	6.061	6.009	0.0091	0.078	C-C	65	6.062	4.827	0.2288	0.003	B-B
30	6.112	6.060	0.0094	0.075	A-A	66	6.110	4.757	0.2518	0.004	A-A
31	6.062	6.008	0.0099	0.071	B-B	67	6.016	4.674	0.2678	0.031	C-C
32	6.112	6.054	0.0104	0.067	A-A	68	6.062	4.509	0.2970	0.002	B-B
33	6.061	6.000	0.0106	0.066	C-C	69	6.107	4.432	0.3230	0.005	A-A
34	6.062	6.001	0.0111	0.063	B-B	70	5.997	4.200	0.3779	0.031	C-C
35	6.112	6.049	0.0112	0.062	A-A	71	6.060	3.874	0.4491	0.002	B-B
36	6.061	5.991	0.0121	0.058	C-C	72	5.890	2.986	0.7541	0.053	A-A

Table 9.6. Tensile test results of Cu16 wt % Zn [110] single crystal
Section AA: $w_0 = 6.014$ mm, $t_0 = 6.012$ mm
Section BB: $w_0 = 6.016$ mm, $t_0 = 6.014$ mm
Section CC: $w_0 = 6.034$ mm, $t_0 = 6.028$ mm

n	w	t	ε	r	sect.	n	w	t	ε	r	sect.
1	6.015	6.013	0.0003	1.000	B-B	34	6.032	5.956	0.0123	0.028	C-C
2	6.033	6.027	0.0003	0.999	C-C	35	6.012	5.940	0.0124	0.028	A-A
3	6.013	6.018	0.0005	0.500	A-A	36	6.014	5.935	0.0136	0.025	B-B
4	6.015	6.011	0.0007	0.333	B-B	37	6.032	5.948	0.0137	0.025	C-C
5	6.013	6.009	0.0007	0.333	A-A	38	6.032	5.940	0.0150	0.023	C-C
6	6.033	6.025	0.0007	0.333	C-C	39	6.014	5.924	0.0154	0.022	B-B
7	6.013	6.008	0.0008	0.250	A-A	40	6.032	5.932	0.0164	0.021	C-C
8	6.033	6.022	0.0012	0.166	C-C	41	6.014	5.912	0.0174	0.019	B-B
9	6.015	6.007	0.0013	0.143	B-B	42	6.032	5.924	0.0177	0.019	C-C
10	6.013	6.005	0.0013	0.143	A-A	43	6.032	5.916	0.0191	0.018	C-C
11	6.012	6.003	0.0018	0.222	A-A	44	6.014	5.900	0.0195	0.017	B-B
12	6.012	6.002	0.0020	0.200	A-A	45	6.014	5.888	0.0215	0.016	B-B
13	6.032	6.010	0.0033	0.111	C-C	46	6.014	5.876	0.0235	0.014	B-B
14	6.014	5.995	0.0035	0.105	B-B	47	6.014	5.864	0.0256	0.013	B-B
15	6.032	6.005	0.0042	0.087	C-C	48	6.014	5.852	0.0276	0.012	B-B
16	6.012	5.986	0.0047	0.077	A-A	49	6.012	5.832	0.0307	0.011	A-A
17	6.014	5.983	0.0055	0.064	B-B	50	6.032	5.804	0.0382	0.009	C-C
18	6.032	5.997	0.0055	0.064	C-C	51	6.014	5.734	0.0480	0.007	B-B
19	6.012	5.978	0.0060	0.059	A-A	52	6.012	5.696	0.0543	0.006	A-A
20	6.012	5.974	0.0067	0.052	A-A	53	6,032	5.692	0.0577	0.006	C-C
21	6.032	5.989	0.0068	0.051	C-C	54	6.014	5.615	0.0690	0.005	B-B
22	6.012	5.972	0.0070	0.050	A-A	55	6.012	5.570	0.0767	0.004	A-A
23	6.012	5.971	0.0072	0.049	A-A	56	6.032	5.580	0.0776	0.004	C-C
24	6.012	5.970	0.0073	0.047	A-A	57	6.014	5.497	0.0902	0.004	B-B
25	6.014	5.971	0.0075	0.046	B-B	58	6.012	4.862	0.2126	0.002	A-A
26	6.012	5.966	0.0080	0.043	A-A	59	6.032	4.824	0.2231	0.001	C-C
27	6.032	5.981	0.0082	0.042	C-C	60	6.012	4.750	0.2366	0.003	B-B
28	6.012	5.964	0.0083	0.041	A-A	61	6.012	3.739	0.4753	0.001	A-A
29	6.012	5.958	0.0094	0.037	A-A	62	6.032	3.624	0.5092	0.001	C-C
30	6.014	5.959	0.0095	0.036	B-B	63	6.012	3.543	0.5291	0.001	A-A
31	6.032	5.973	0.0095	0.036	C-C	64	6.010	3.448	0.5596	0.002	B-B
32	6.032	5.964	0.0110	0.031	C-C	65	6.032	3.250	0.6181	0.001	C-C
33	6.014	5.947	0.0115	0.038	B-B	66	6.008	2.711	0.7981	0.002	B-B

Table 9.7. Tensile test results of Cu30 wt % Zn [110] single crystal
Section AA: w_0 = 6.024 mm, t_0 = 6.028 mm
Section BB: w_0 = 5.988 mm, t_0 = 5.995 mm
Section CC: w_0 = 6.003 mm, t_0 = 6.005 mm

n	w	t	ε	r	sect.	n	w	t	ε	r	sect.
1	5.987	5.994	0.0003	1.001	B-B	39	6.023	5.992	0.0062	0.028	A-A
2	6.002	6.004	0.0003	1.000	C-C	40	6.023	5.984	0.0075	0.023	A-A
3	5.987	5.993	0.0005	0.501	B-B	41	5.987	5.932	0.0107	0.016	B-B
4	6.023	6.026	0.0005	0.500	A-A	42	6.023	5.928	0.0169	0.010	A-A
5	6.002	6.003	0.0005	0.500	C-C	43	6.002	5.899	0.0180	0.009	C-C
6	5.987	5.992	0.0007	0.334	B-B	44	5.987	5.883	0.0190	0.009	B-B
7	6.023	6.025	0.0007	0.333	A-A	45	6.022	5.904	0.0211	0.016	A-A
8	6.023	6.025	0.0007	0.333	A-A	46	6.022	5.893	0.0230	0.015	A-A
9	6.002	6.002	0.0007	0.333	C-C	47	5.985	5.859	0.0234	0.022	B-B
10	5.987	5.991	0.0008	0.250	B-B	48	6.022	5.890	0.0235	0.014	A-A
11	6.023	6.024	0.0008	0.250	A-A	49	6.022	5.868	0.0272	0.012	A-A
12	6.002	6.001	0.0008	0.250	C-C	50	6.002	5.821	0.0313	0.005	C-C
13	6.023	6.023	0.0010	0.200	A-A	51	6.002	5.820	0.0315	0.005	C-C
14	5.987	5.990	0.0010	0.200	B-B	52	6.002	5.819	0.0316	0.005	C-C
15	6.023	6.023	0.0010	0.200	A-A	53	6.001	5.818	0.0320	0.011	C-C
16	6.002	6.000	0.0010	0.200	C-C	54	5.985	5.807	0.0324	0.016	B-B
17	5.987	5.989	0.0012	0.167	B-B	55	5.985	5.797	0.0341	0.015	B-B
18	6.023	6.022	0.0012	0.167	A-A	56	5.984	5.778	0.0375	0.018	B-B
19	6.002	5.999	0.0012	0.167	C-C	57	5.983	5.754	0.0419	0.020	B-B
20	5.987	5.988	0.0013	0.143	B-B	58	5.983	5.734	0.0453	0.019	B-B
21	6.002	5.998	0.0013	0.143	C-C	59	6.000	5.664	0.0590	0.009	C-C
22	5.987	5.987	0.0015	0.125	B-B	60	5.980	5.636	0.0631	0.022	B-B
23	6.002	5.997	0.0015	0.125	C-C	61	6.016	5.598	0.0753	0.018	A-A
24	6.023	6.019	0.0017	0.111	A-A	62	6.012	5.522	0.0897	0.023	A-A
25	5.987	5.986	0.0017	0.111	B-B	63	5.991	5.411	0.1062	0.019	C-C
26	6.002	5.996	0.0017	0.111	C-C	64	5.999	5.329	0.1274	0.034	A-A
27	5.987	5.985	0.0018	0.100	B-B	65	5.916	5.337	0.1284	0.104	B-B
28	6.023	6.018	0.0018	0.100	A-A	66	5.978	5.093	0.1689	0.025	C-C
29	6.002	5.995	0.0018	0.100	C-C	67	5.963	5.010	0.1952	0.055	A-A
30	6.023	6.017	0.0020	0.091	A-A	68	5.886	4.968	0.2051	0.091	B-B
31	5.987	5.984	0.0020	0.091	B-B	69	5.960	4.774	0.2366	0.031	C-C
32	6.002	5.994	0.0020	0.091	C-C	70	5.950	4.729	0.2551	0.051	A-A
33	6.023	6.016	0.0022	0.083	A-A	71	5.942	4.660	0.2638	0.040	C-C
34	5.987	5.980	0.0027	0.067	B-B	72	5.901	4.451	0.3166	0.057	C-C
35	6.023	6.012	0.0028	0.062	A-A	73	5.823	3.615	0.5379	0.060	C-C
36	5.987	5.977	0.0032	0.056	B-B	74	5.876	3.559	0.5403	0.036	B-B
37	6.002	5.978	0.0047	0.037	C-C	75	5.894	3.560	0.5485	0.041	A-A
38	6.002	5.978	0.0047	0.037	C-C						

9.8   Instability of the strain ratio and crystallographic orientation in strained real *f.c.c.* single crystals

The data in Table 9.3 illustrate the effect of the stacking fault energy on the value of the instability index in single crystals of *f.c.c.* metals (aluminium [9.26], copper [9.27], silver [9.28], and four brasses [9.19] The decrease of the stacking fault energy induces an increase of the index φ, hence a decrease of the stability of the strain ratio r; the plastic anisotropy coefficient (and thereby the crystallographic orientation) is more stable during the uniaxial deformation of copper than in the case of silver or brasses.

Instability of crystallographic orientation in imperfect single crystals of *f.c.c.* metals is determined by factors defining their deviation from the extreme state. In the present research it has been expressed by the coefficients α and ρ. It is obvious that the characteristics of the instability of crystallographic orientation of deformed single crystals is the basis for the investigations of the evaluation of the mechanism of deformation texture in polycrystalline metals.

However, the main advantage of the author's method described in Chapter 9 is its relative simplicity; the quantitative evaluation of the change of the strain ratio in the range of small strains (instability at the tensile test) is not too much complicated, while the direct expression of the instability of crystallographic orientation is much more complex.

9.9   References

9.1. A.Modrzejewski, *Estimate of Micromosaics of Large Copper Single Crystal Plates*, Matériaux et Techniques, VIII-IX (1985) E 123.

9.2. W.Truszkowski and J.Bonarski, *On the Imperfection of Crystallographic Orientation in FCC Single Crystals*, Z. Metallkd., **91**, (2000), in print.

9.3. W.Truszkowski and A.Modrzejewski, *Influence of Stacking Fault Energy on Instability of Crystallographic Orientation in Tensile Tested Brass Single Crystals*, Arch. of Metallurgy, **35**, 219 (1990).

9.4. R.Dupke, and W.Reimers, *X-ray Diffraction Investigations on Individual Grains in the Polycrystalline Ni-base Superalloy IN 939 During Cyclic Loading I: X-ray Rocking Curve Broadening*, Z. Metallkd, **86**, 371 (1995).

9.5. W.M.Baldwin Jr., *Effect of Rolling and Annealing upon the Crystallography*, Metallography and Physical Properties of Copper Strip, Trans. AIME, **166**, 591 (1946).

9.6. W.T.Roberts, *Crystallographic Aspects of Directionality in Sheet*, Sheet Met. Ind., **39**, 855 (1962).

9.7. W.Truszkowski, S.Wierzbiński, *Izmenene koefficienta plastičeskoj anizotropii monokristallov medi s orientacjej blizkoj [001] pri rastjazenii*, Fizika Metallov i Metallovedenie, **56**, 1195 (1983).

9.8. W.Truszkowski, S.Wierzbiński, A.Modrzejewski, J.Baczyński, G.S.Burhanov, I.V.Burov, and O.D.Čistjakov, *Influence of Deviation from <001>, <011> and <111> Orientations on the Variation of Strain Ratio in Deformed Nickel Single Crystals*, Arch. of Metallurgy, **32**, 165 (1987).

9.9. T.Takeuchi, *Orientation Dependence of Work Hardening of Copper Single Crystals Near the [001] Axis*, J. Phys. Soc. Japan, **40**, 741 (1976).

9.10. W.Truszkowski, S.Wierzbiński and A.Modrzejewski, *Influence of Mosaic Structure on Instability of the Strain Ratio in Deformed Copper Single Crystals*, Bull. Acad. Pol. Sci., sér. techn., **30**, 367 (1982).

9.11. J.R.Schneider, *Diffraction Properties of a Copper Mosaic Crystal: Model Calculations based on γ-ray Diffractometry*, J.Appl. Cryst., **8**, 195 (1975).

9.12. W.Truszkowski, *The Impact of Texture in Single crystals of FCC Metals on Mechanical Behaviour and Instability of Orientation*, Proc. Eight Intern. Conf. on Textures of Materials (ICOTOM 8), Ed. by Kallend and Gottstein, The Met. Soc., 537 (1988).

9.13. M.Ya. Skorokhod, *Establishing the Depth and Structure of Damaged Surface Layers of Monocrystals by X-Ray Topography*, Met. Phys. Adv. Tech., **16**, 661 (1997).

9.14. J. Bonarski, *Computer Programmes*, (1997) - unpublished.

9.15. W.Truszkowski et J.Kloch, *Modifications du coefficient d'anisotropie plastique en cours de déformation hétérogène*, Matériaux et Techniques, E 17 (1985).

9.16. J.Kloch, W.Truszkowski, *The Method of Determination of the Fitting Function Based on Maximal Errors*, Bull. Pol. Ac.: Techn., **34**, 683 (1986).

9.17. W.Truszkowski, J.Kloch, *Application of the Maximal Error Method for the Calculation of the r(ε) Function*, Bull. Pol. Ac.: Techn., **34**, 691 (1986).

9.18. W.Truszkowski and J.Kloch, *The Variation of Strain Ratio at the Tensile Test Described by a Hyperbolic Function*, Textures and Microstructures, **26 – 27**, 531 (1996).

9.19. W.Truszkowski, A.Modrzejewski and J.Baczyński, *Variation of the Strain Ratio in Tensile Tested <011> Brass Single Crystals*, Bull. Pol. Ac.: Techn., **37**, 471 (1989).

9.20. P.Coulomb, *Estimation expérimentale des énergies de défaut d'empilement et de paroi d'antiphase*, J. Micros. Spectrosc. Electron., **3**, 295 (1978).

9.21. P.R.Thornton, T.E.Mitchell and P.B.Hirsch, *The Dependence of Cross-slip on Stacking-fault Energy in Face-centered Cubic Metals and Alloys*, Phil. Mag., **7**, 1349 (1962).

9.22. L.B.Brown, *The Self-stress of Dislocations and the Shape of Extended Nodes*, Phil. Mag., **10**, 441 (1964).

9.23. J.Hedworth and G.Pollard, *The Influence of Stacking-Fault Energy on the Creep Behaviour of α-Brasses*, Met. Sci. Journ., **5**, 41 (1971).

9.24. Ja.D.Višnjakov, *Defekty upakovki v kristalličeskoj strukture*, Metallurgia, Moskva 1970.

9.25. P.C.J.Gallagher, *The Influence of Alloying, Temperature, and Related Effects on the Stacking Fault Energy*, Metall. Trans., **1**, 2429 (1970).

9.26. W.Truszkowski, A.Piątkowski and J.Kloch, *Plastic Anisotropy in Strained Aluminium [110] Single Crystals*, unpublished.

9.27. W.Truszkowski, J.Gryziecki and J.Jarominek, *Variation of Strain Ratio in Cube Plane of Copper*, Metals Technology, **6**, 439 (1979).

9.28. W.Truszkowski, J.Gryziecki and J.Jarominek, *Variation of Plastic Strain Ratio in the {001} Crystallographic Plane of Silver*, Bull. Pol. Ac.: Tech., **31**, 31 (1983).

Chapter 10

The multifarious application of the $r(\varepsilon)$ function in research and technology

Plastic anisotropy and its transition in the deformation process of metals is expressed by the function $r(\varepsilon)$ describing the change of the strain ratio in a static tensile test. It illustrates the character of changes in the directionality of mechanical properties, in particular those, which are evidence of the plasticity of the material. The increase or decrease in the value of the coefficient r is, first of all, due to the rotation of crystallographic lattice and the newly forming tensile texture. The texture, which is chiefly responsible for the directionality of mechanical properties in annealed, single-phase materials, is being changed, whereas in multiphase materials the anisotropy is modified also by the transformations of the form of non-metallic inclusions and of the particular phases, as well as by internal stresses.

The studies of plastic anisotropy in mono- and polycrystalline materials, carried out for many years, have shown that the function $r(\varepsilon)$ properly determined on the basis of experimental data and rationally interpreted, represents a compendium of the knowledge of the directional properties of metals and their applicability to the process of plastic working.

When measuring the coefficient r, the sample is strained to considerable elongation, usually up to the onset of instability (or even to the sample rupture), which often produces a radical change of the texture and anisotropy and consequently, the value of the coefficient, obtained in this way, cannot be regarded as corresponding to the looked-for characteristics. It may appear that this difficulty can be overcome by reducing the elongation at which the measurement is carried out. However, it should be noted that – as it has been indicated in many studies by the author – the measuring error is here, in general, so great that the determined value of the anisotropy coefficient is not relevant. In this situation, the only solution of the problem of defining the characteristics of plastic anisotropy appears the extrapolation of the function

$r(\varepsilon)$ to the initial state, i.e. to $\varepsilon = 0$. To do this it is necessary to express the dependence $(r - \varepsilon)$ by a mathematical function. In this field, the extensive research studies of the author and his co-workers, analyzing the change of anisotropy coefficient with elongation, both in single crystal of *f.c.c.* metals and in polycrystalline cubic metals, enabled to define the dependence $(r - \varepsilon)$ by a hyperbolic function (Eq. 5.1) with five constant parameters. Equation (5.1) describes each case of the $r(\varepsilon)$ function, found in practice: from the arbitrary position of asymptotes on the $\varepsilon - r$ plane to a straight line, when the a_3 parameter is equal to zero. The advantage of the proposed method is that the determination of the approximating function can be based, to a considerable extent, on the quantities (ε_i, r_i) measured at large elongation, where the experimental error in a correctly performed experiment is rather small.

The coefficient $r_0 = r(0)$, obtained in this way, is an essential property and it can be also determined indirectly, starting from the crystallographic orientation in single crystals or the quantitative description of texture in polycrystalline metals.

Only for metallic single crystals with stable crystallographic orientation and for those polycrystalline aggregates in which the static tension test allows to state that $r(\varepsilon) = \text{const.} = r_0$, it can be assumed that the value r_0, determined on the sample strained up to the limit of uniform elongation, is an essential property of the material. Thus, it seems strange that authors of numerous studies often attempt to refer the value of r_a (corresponding to large elongation) to the anisotropy coefficient, obtained in an indirect way, from the physical characteristics of an undeformed material. Such a method (i.e. based on the assumption $r_a = r_0$) could be applied e.g. in case of mild steel, but only after establishing of the stability of the $r(\varepsilon)$ function.

It seems that basing on the results of studies on instability of the strain ratio and its relation to the stacking fault energy (Chapter 9), it might be possible to characterize the physical properties of the material at which the determined at considerable strain the r_a value, approaches with sufficient accuracy the intrinsic property (r_0) of the tested sample.

In the technology of plastic working of metals, plastic anisotropy is considered as one of the factors (besides the work hardening exponent n), defining the deep drawability of metal sheets; and here the index r_a, which defines the anisotropy of the metal after marked deformation, may be of use in the technological process. However, it should be remembered that the anisotropy coefficient is determined in a test of uniaxial deformation, while in the process of drawing thin metal sheets we have to do with biaxial straining.

Taking into account the conclusions of Chapters 4, 6 and 7, concerning the physical sense of the r_0 coefficient as an intrinsic property of the material, it is to be stressed that the main problem dealt with in the monograph is the $r(\varepsilon)$ function describing the variation of the strain ratio with strain at the uniaxial tensile test, both (1) in the zone of small strains, as well as (2) at large deformation.

One would suggest that (1) when searching for the physical characteristics of plastic anisotropy (not changed under the influence of strain at the tensile test) the zone of very small deformation should be considered. However, it should be remarked that: 1° there exist a very narrow „forbidden zone" (Sect. 6.3.2) in the $r - \varepsilon$ relation of Δw and Δt values smaller than c_0 (i.e. of the order of 10^{-3} mm) where the calculated r values have no physical meaning, 2° at small ε values the experimental error of r values is sometimes exceptionally large (Sect. 6.3.2), 3° there exists, in the $r - \varepsilon$ diagram, a small zone of adjustment (depending on the elastic characteristics of the material, the precision of the sample preparation, testing machine clearance ...) in which the obtained (ε_i, r_i) data are not relevant. Therefore, contrary to the first suggestion, the description of the $r(\varepsilon)$ equation from experimental data should be based less on the first but much more on the medium zone of strain where the measurement error is relatively small.

An essential support for the r_0 coefficient as a material's property are the results on texture and anisotropy in annealed single phase metals and alloys (Chapter 4.1); the one-to-one relation between texture and plastic anisotropy was observed only in the case when r_0 (and not r_a – calculated at the limit of uniform elongation) was used in the calculations.

Similarly, the research on *f.c.c.* single crystals characterized by the stable behaviour in the tensile test (Chapter 7) made it possible to confirm the r_0 value as an intrinsic material's property: confirmation of the theoretical values of parameters of the $r(\varepsilon)$ function by the tensile test results in „perfect single crystals".

On the other hand (2), the value of the r_a (strain ratio calculated at a considerable strain, often corresponding to the limit of uniform elongation) is of great importance for technological purposes. In the technology of plastic working of metals plastic anisotropy is considered, besides the work-hardening exponent, as one of the most important factors influencing the deep drawability of metal sheets. In the technology of working of metal sheets it is important to have to one's disposal a method of evaluation of the properties reproducing, as closely as possible, the deformation mechanism in the industry. This requirement is satisfied in the tensile test leading up to the critical state, i.e. up to the instability limit.

Other original methods, described in the monograph, e.g. the method of approximating the experimental results on the basis of maximal errors, which proved to be useful in the analysis of the $r(\varepsilon)$ function, may find application in the case when the great experimental scatter around the examined function is asymmetric and is varying with elongation; in the case of the function $r(\varepsilon)$ the scatter is extremely great in the range of small ε values and small – when the experiment is carried out correctly – in the case of considerable elongations. A method of quantitative estimation of the asymmetry of the experimental scatter and a way of its radical reduction have also been proposed. A method, tested with success on the example of the $r(\varepsilon)$ function, can be applied at any place, where this phenomenon occurs.

Table 10.1. Tensile test results of Al-4 [110] single crystal
Section AA: w_0 = 5.917 mm, t_0 = 5.918 mm
Section BB: w_0 = 5.908 mm, t_0 = 5.919 mm
Section CC: w_0 = 5.897 mm, t_0 = 5.914 mm

	AA		BB		CC	
	ε	r	ε	r	ε	r
1	0.0012	0.7500	0.0011	0.7513	0.0037	0.5727
2	0.0015	0.8000	0.0018	0.5723	0.0040	0.7161
3	0.0029	0.3075	0.0025	0.6677	0.0042	0.7877
4	0.0184	0.1126	0.0033	0.6676	0.0137	0.2653
5	0.0253	0.0716	0.0052	0.6323	0.0120	0.1091
6	0.0301	0.0987	0.0096	0.6771	0.0311	0.1433
7	0.0377	0.0877	0.0308	0.1031	0.0477	0.1378
8	0.0442	0.1106	0.0423	0.0918	0.0613	0.1212
9	0.0546	0.0803	0.0516	0.1053	0.0720	0.1159
10	0.0637	0.0835	0.0599	0.1100	0.0830	0.1351
11	0.0731	0.0667	0.0703	0.1342	0.0985	0.1223
12	0.0824	0.0703	0.0827	0.0993	0.1050	0.1139
13	0.1129	0.2331	0.0886	0.1107	0.1149	0.1012
14	0.1107	0.1243	0.1010	0.0898	0.1306	0.1005
15	0.1189	0.0477	0.1150	0.0745	0.1298	0.1360
16	0.1345	0.0789	0.1116	0.3401	0.1451	0.1732
17	0.1475	0.0571	0.1301	0.1687	0.1665	0.1234
18	0.1638	0.0968	0.1545	0.0848	0.1735	0.1254
19	0.1787	0.0604	0.1592	0.1040	0.1869	0.1097
20	0.1918	0.0492	0.1733	0.0795	0.2039	0.1513
21	0.2059	0.0367	0.1854	0.1256	0.2244	0.1074
22	0.2403	0.0889	0.2112	0.0908	0.2403	0.0996
23	0.2574	0.0573	0.2250	0.0749	0.2579	0.1093
24	0.2838	0.0516	0.2425	0.0815	0.2875	0.0984
25	0.3152	0.0451	0.2734	0.0745	0.3102	0.1110
26	0.3490	0.0557	0.2965	0.0777	0.3347	0.0932
27	0.4043	0.0687	0.3258	0.0618	0.3626	0.0883
28	0.4417	0.0652	0.3611	0.0667	0.4066	0.1005
29	0.5054	0.0682	0.3890	0.0687	0.4692	0.1072
30	0.5790	0.0759	0.4523	0.0769	0.5210	0.1134
31	0.5978	0.0792	0.5020	0.0846	0.6188	0.1115
32			0.5915	0.0929	0.7462	0.1353
33			0.6927	0.1145	0.8555	0.1647
34			0.7259	0.1240		

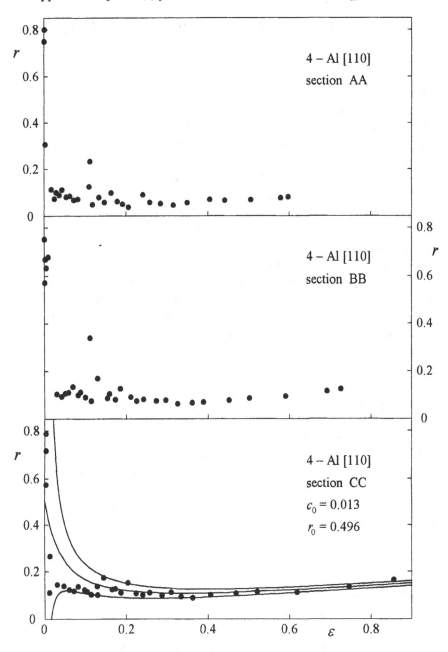

Figure 10.1. r(ε) relations for three sections of the aluminium Al-4 [110] single crystal [10.3]

The results obtained in the author's investigations of the physical sense of the anisotropy coefficient and of the causes of instability of crystallographic orientation in nominally stable single crystals offer the prospects for a quantitative evaluation of the experimental error. On the other hand, the problem of instability of crystallographic orientation in deformed single crystals is at the bases of the mechanism of deformation texture in polycrystalline metals.

Model investigations of perfect single crystals and experimental studies of real *f.c.c.* crystals enabled to assess the imperfection of their crystallographic orientation, and this, as a consequence, leads to the criterion of the "single crystal texture". This concept signaled by the author for some time now [10.2], has not found as yet any marked response in the world literature, in spite of its evident influence on the instability of crystallographic orientation in the deformed crystals. Maybe, the hardly discernible direct effect of the imperfection of crystallographic orientation on the mechanical properties is responsible for the rather small interest in the single crystal texture.

Finally, it is worth mentioning that the proposed method of calculation of the parameters of the $r(\varepsilon)$ function may be very useful in the case of aluminium single crystal cylindrical samples. The difficulty of setting in motion a sufficient number of slip systems at the onset of deformation (aluminium is characterized by a high stacking fault energy) leads to irregular form of the strained sample (its cross-section is no more circular), and consequently makes impossible the measurement of Δw and Δt values. It has been shown that in this case the first part of the $r(\varepsilon)$ relation can be calculated by utilizing the medium zone of strain.

A round section sample of aluminium Al-4 [110] single crystal was subjected to tensile test [10.3]. The procedure of the examination was described in Chapter 7. The diameter of the sample was 6 mm, its length ca 70 mm; the crystallographic parameters: $\alpha = 13'$, $\rho = 16'$. The change in the dimensions of the strained sample was observed in three sections: AA, BB (in the middle) and CC. The results are shown in Table 10.1 and in Fig. 10.1. The scarcity of credible data in sections AA and BB within the zone of small strain makes impossible the calculation of the $r(\varepsilon)$ function. Whereas in section CC even a limited number of experimental points made possible the calculation of the function parameters, and the calculated r_0 (for the section CC) was very close to the theoretical value $r_{0\,[110]\,cal} = 0.495$ (the theoretical value is $r_{0\,[110]\,theor} = 0.500$). The above described results show that in case of high stacking fault energy single crystals, even the application of the maximal error method combined with the hyperbolic description makes difficult the calculation of the $r(\varepsilon)$ function parameters.

The final conclusion. The author is aware of the fact that certain phenomena, described in the monograph, to which also a quantitative approach has been proposed, are not often observed outside the area explored in this study. It seems, however, that the developed methods should be of interest to the reader, since when extrapolated to other research areas, they might extend and increase the effectiveness of the scientific apparatus.

This refers to such solutions, proposed by the author, as the approximation of the experimental results by the method of the maximal error in the case of its considerable and asymmetric scatter, quantitative approach to the experimental error, the method of the extrapolation of a function beyond the range in which the measurements were carried out, when the possibility of the assessment of error is limited, quantitative description of the instability of crystallographic orientation in an indirect way, on the basis of the observed change in the anisotropy of the mechanical properties.

10.1 References

10.1. W.Truszkowski and J.Kloch, *The Variation of Strain Ratio at the Tensile Test Described by a Hyperbolic Function*, Textures and Microstructures, **26-27**, 531 (1996).

10.2. W.Truszkowski, *The Impact of Texture in Single Crystals of FCC Metals on Mechanical Behaviour and Instability of Orientation*, Proc. Eight Intern. Conf. on Textures of Marerials (ICOTOM 8), Ed. by Kallend and Gottstein, The Met. Soc., 537 (1988).

10.3. W.Truszkowski, A.Piątkowski and J.Kloch, *Plastic Anisotropy in Strained Aluminium [110] Single Crystals*, unpublished.

Frequently used symbols and abbreviations

r — strain ratio, coefficient of plastic anisotropy

r_0 — strain ratio at zero strain

$r_{0\ theor}$ — theoretically calculated r_0 value (in single crystal)

$r_{0\ cal}$ — calculated value $r_{0\ cal} = r(0)$ (in single crystal)

r_a — strain ratio at the limit of uniform elongation

ε — longitudinal natural plastic strain

ε_w — natural plastic strain in the width direction

ε_t — natural plastic strain in the thickness direction

ε_0 — initial plastic strain in Swift formula

D — plastic anisotropy coefficient characterized by symmetrical experimental scatter

D_0 — plastic anisotropy coefficient calculated from r_0 values

K — plastic anisotropy coefficient (Krupkowski)

K_0 — plastic anisotropy coefficient at zero strain

j — plastic anisotropy coefficient (Jegaden)

j_0 — plastic anisotropy coefficient at zero strain

n — strain hardening exponent in Ludwik formula

m — strain hardening exponent in Krupkowski and Swift formulae

z — cold work, reduction in area

z_1 — initial deformation in Krupkowski formula

σ — true stress in the tensile test

σ_0 — initial stress in the Ludwik formula

w_0 — initial width of the cross-section of the tensile test sample

t_0 — initial thickness of the cross-section of the tensile test sample

w — width of the cross-section of the tensile tested sample

t	–	thickness of the cross-section of the tensile tested sample
c_0	–	maximum experimental error at the onset of the tensile test
c	–	maximum experimental error during the tensile test
A_p	–	index of plastic anisotropy calculated from r_0 values
$(A_p)_a$	–	index of plastic anisotropy calculated from r_a values
T_c	–	index of texture (Truszkowski and Król)
J	–	index of texture (Sturcken and Croach)
T_c	–	index of cube texture
T_{rol}	–	index of rolling texture
AA,BB,CC	–	sections of the tensile test sample (along the length)
RD	–	rolling direction
ND	–	normal direction
TD	–	transverse direction
A_r	–	asymmetry index calculated from r values
A_D	–	asymmetry index calculated from D values
φ	–	instability index
γ	–	stacking fault energy
G	–	shear modulus
b	–	Burgers vector
α, θ	–	angle between [u v w] nominal direction in a single crystal and the sample axis
[u v w]	–	Miller index of crystallographic direction
ρ	–	half-width of the rocking curve
FWHM	–	full width at half maximum of intensity curve
DSL	–	damaged surface layer